LIST OF TITLES

Already published

A Biochemical Approach to Nutrition	R.A. Freedland, S. Briggs
Biochemical Genetics (second edition)	R.A. Woods
Biological Energy Conservation (second edition)	C.W. Jones
Biomechanics	R.McN. Alexander
Brain Biochemistry (second edition)	H.S. Bachelard
Cellular Degradative Processes	R.T. Dean
Cellular Development	D.R. Garrod
Cellular Recognition	M.F. Greaves
Control of Enzyme Activity	P. Cohen
Cytogenetics of Man and other Animals	A. McDermott
Enzyme Kinetics: The Steady-State Approach (second edition)	P.C. Engel
Functions of Biological Membranes	M. Davies
Genetic Engineering: Cloning DNA	D. Glover
Hormone Action	A. Malkinson
Human Evolution	B.A. Wood
Human Genetics	J.H. Edwards
Immunochemistry	M.W. Steward
Insect Biochemistry	H.H. Rees
Isoenzymes	C.C. Rider, C.B. Taylor
Membrane Biochemistry	E. Sim
Metabolic Regulation	R. Denton, C.I. Pogson
Metals in Biochemistry	P.M. Harrison, R. Hoare
Molecular Virology	T.H. Pennington, D.A. Ritchie
Motility of Living Cells	P. Cappuccinelli
Muscle Contraction	C.R. Bagshaw
Plant Cytogenetics	D.M. Moore
Polysaccharide Shapes	D.A. Rees
Population Genetics	L.M. Cook
Protein Biosynthesis	A.E. Smith
RNA Biosynthesis	R.H. Burdon
The Selectivity of Drugs	A. Albert
Transport Phenomena in Plants	D.A. Baker

Editors' Foreword

The student of biological science in his final years as an undergraduate and his first years as a graduate is expected to gain some familiarity with current research at the frontiers of his discipline. New research work is published in a perplexing diversity of publications and is inevitably concerned with the minutiae of the subject. The sheer number of research journals and papers also causes confusion and difficulties of assimilation. Review articles usually presuppose a background knowledge of the field and are inevitably rather restricted in scope. There is thus a need for short but authoritative introductions to those areas of modern biological research which are either not dealt with in standard introductory textbooks or are not dealt with in sufficient detail to enable the student to go on from them to read scholarly reviews with profit. This series of books is designed to satisfy this need. The authors have been asked to produce a brief outline of their subject assuming that their readers will have read and remembered much of a standard introductory textbook of biology. This outline then sets out to provide by building on this basis, the conceptual framework within which modern research work is progressing and aims to give the reader an indication of the problems, both conceptual and practical, which must be overcome if progress is to be maintained. We hope that students will go on to read the more detailed reviews and articles to which reference is made with a greater insight and understanding of how they fit into the overall scheme of modern research effort and may thus be helped to choose where to make their own contribution to this effort. These books are guidebooks, not textbooks. Modern research pays scant regard for the academic divisions into which biological teaching and introductory textbooks must, to a certain extent, be divided. We have thus concentrated in this series on providing guides to those areas which fall between, or which involve, several different academic disciplines. It is here that the gap between the textbook and the research paper is widest and where the need for guidance is greatest. In so doing we hope to have extended or supplemented but not supplanted main texts, and to have given students assistance in seeing how modern biological research is progressing, while at the same time providing a foundation for self help in the achievement of successful examination results.

Glycoproteins

R.C. Hughes

Medical Research Council National Institute for Medical Research
Mill Hill, London

London New York

Chapman and Hall

First published in 1983 by
Chapman and Hall Ltd
11 New Fetter Lane
London EC4P 4EE
Published in the USA by
Chapman and Hall
733 Third Avenue
New York NY 10017

Printed in Great Britain by
J.W. Arrowsmith Ltd, Bristol

ISBN 0 412 24150 1

British Library Cataloguing in Publication Data

Hughes, R.C.
Glycoproteins.—(Outline studies in biology)
1. Glycoproteins 2. Proteoglycans
3. Biological chemistry
I. Title II. Series
574.19′245 QP552.G5

ISBN 0-412-24150-1

Library of Congress Cataloging in Publication Data

Hughes, R. Colin.
Glycoproteins.

(Outline studies in biology)
Bibliography: p.
Includes index.
1. Glycoproteins. I. Title. II. Series: Outline studies in biology (Chapman and Hall([DNLM:
1. Glycoproteins. QU 55 H894g]
QP552.G59H826 1983 574.19′2454 82-14593
ISBN 0-412-24150-1

Contents

Acknowledgements

I thank the following friends and colleagues for many helpful comments during the preparation of this book: W. Carter, M. Fukuda, P. Gleeson, H. Schachter and N. Sharon. They can be credited with many improvements but I take full responsibility for the final produce.

1 Introduction

Glycoproteins are proteins containing oligosaccharides covalently attached to selected aminoacid residues. They have a long history and are believed to have important biological functions [1–3], a view that I hope is reinforced by this book.

The term oligosaccharide is commonly used for carbohydrate polymers comprised of from two to ten monosaccharide residues. There is an enormous variety of monosaccharides in nature, the most common sugar being D-glucose. Fig. 1.1 shows three formulae for D-glucose. Monosaccharides differ in the length of the carbon chain, the number of hydroxyl groups or equivalent groups, e.g. amino and the disposition of hydroxyls and hydrogens about the central axis. The open chain 6-carbon structure (a) is a theoretical formula only; in reality the monosaccharides exist as cyclic forms with either five or six atoms (one of which is oxygen) in the ring. Each monosaccharide exists usually in one particular ring-form, e.g. a six-membered *pyranose* for glucose (b), and in solution adopts a preferred conformation e.g. a *chair conformation* (c). Very little is yet known about the conformations adopted by the monosaccharide units in glycoproteins but it is important to bear in mind that there is a three dimensionality in the carbohydrate sequences of glycoproteins that must ultimately be considered to understand fully the structure, metabolism and function of these molecules [4]. In oligosaccharides the cyclic monosaccharides are joined together by covalent bonds in which the hydroxyl of carbon atom 1 is reacted with any available hydroxyl other than C(1) of a second monosaccharide with the elimination of a water molecule. The reaction is exactly analogous to the formation of a glycoside, e.g. a methyl (Me) glycoside and obviously can be multiplied to engage any number of monosaccharide units. Since the configuration at C(1) can vary α- or β-glycosidic linkages can be formed.

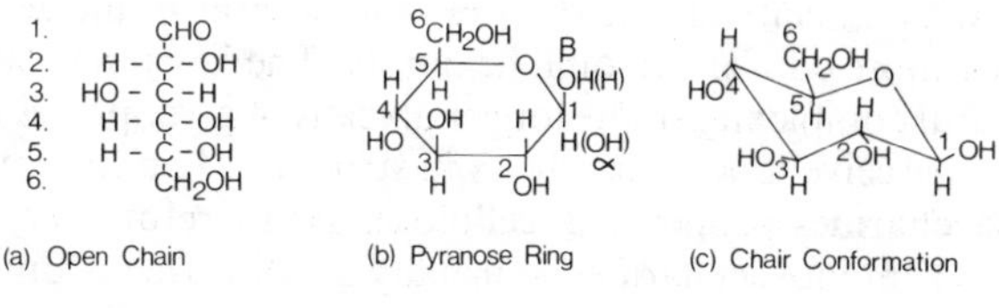

(a) Open Chain (b) Pyranose Ring (c) Chair Conformation

Fig. 1.1 Representations of glucose.

Table 1.1 Glycoproteins, proteoglycans and glycolipids

Commonly found monosaccharide constituents	*Additional constituents*		*Location*
	Common	*Occasional*	
Glycoproteins			
Glucose Galactose Mannose Glucosamine* Galactosamine* Neuraminic acid* Fucose	Protein core	Phosphate Sulphate	Secretions Cell membranes Extracellular matrix and connective tissue
Proteoglycans			
Galactose Xylose Glucuronic acid Iduronic acid Glucosamine* Galactosamine*	Sulphate Protein core		Extracellular matrix and connective tissue
Glycolipids			
Glucose Galactose Glucosamine* Galactosamine* Neuraminic acid* Fucose	Ceramide	Sulphate	Cell membranes Minor plasma constituents

The following abbreviations will be used: glucose (glc, G), mannose (man, M), fucose (fuc, F); galactose (gal, Ga); *N*-acetylglucosamine (glcNAc, Gn); *N*-acetylgalactosamine (galNAc, Gan); *N*-acetylneuraminic acid (neuNAc, Sa, N)
* Always *N*-acylated

As will be seen, the carbohydrate chains of glycoproteins typically range from disaccharides to very complex structures containing eighteen monosaccharides or even more. Strictly speaking, therefore, the longer carbohydrate chains should be classified as polysaccharides or *glycans* and the latter term is often used, for example we refer to glycans of ovalbumin. While disaccharide units exist in glycoproteins such as collagen or submaxillary gland secretions, other carbohydrate chains or glycans of glycoproteins consist of up to seven different monosaccharides (Table 1.1). The glycans are highly branched in a tree-like structure with a relatively rigid trunk and several branches that may have rather more conformational flexibility. The key point, however, is the structural complexity of the longer glycans of glycoproteins: it is not possible to perceive much periodicity in structure. The circumstances in the polysaccharides proper, e.g. cellulose, are therefore very different. The range of monosaccharides is usually small (often a unique sugar such as glucose in cellulose) and an underlying repeating sequence of monosaccharides joined to each other by characteristic glycosidic

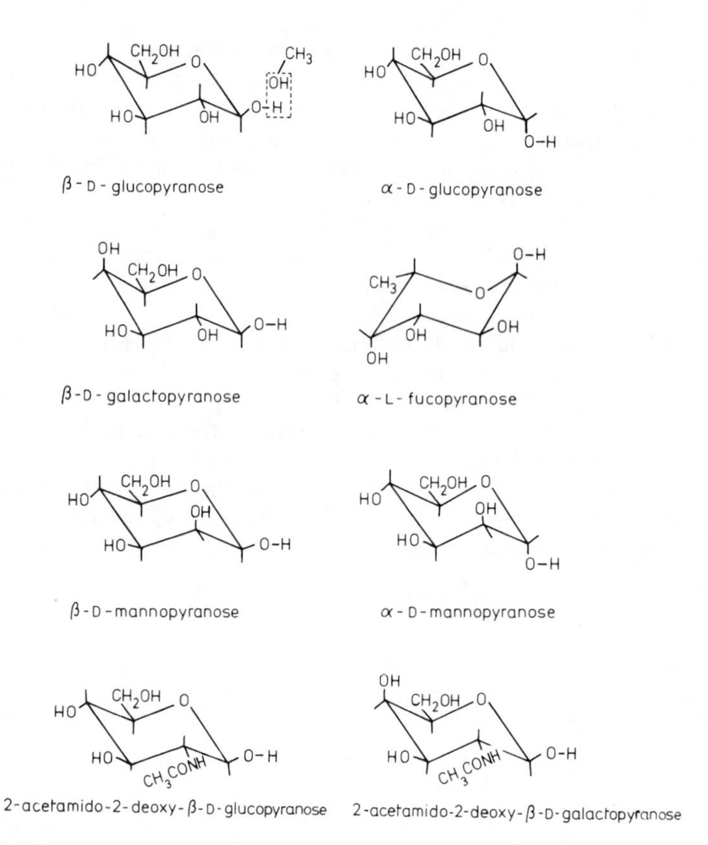

Fig. 1.2 The ring structures of monosaccharides present in glycoproteins. Hydrogen substituents are omitted for clarity. See Fig. 2.3 for the structure of neuraminic acids. The formation of a (methyl) glycoside is shown for β-D-glucopyranose.

linkages, e.g. $\beta 1 \rightarrow 4$ in cellulose can be readily seen. These glycans are also much longer usually than in the glycoproteins. Nevertheless these rather artificial divisions between polysaccharides 'proper' and the carbohydrate chains of glycoproteins are being broken down. In the first place it is now clear that what were once considered as 'pure' polysaccharides such as starch, glycogen or heparin with typical repeating structures are in fact synthesized on a protein core: they start life as glycoproteins. Other examples of this type are the acidic polysaccharides of connective tissues such as the chondroitin sulphates, dermatan sulphate and heparan sulphate. All of these polysaccharides are firmly linked to protein cores and also qualify as a special type of glycoprotein. The convention is to use a term *proteoglycan* for these substances of connective tissue and this convention will be followed in this book. Finally, it should be mentioned for completeness, although we shall not be considering them in great detail, oligosaccharides similar to those found in glycoproteins may be substituted onto the lipid ceramide. These specific *glycolipids* are very important constituents of cell membranes. Often the functions and indeed the metabolism of the ceramide sugars are analogous to similar oligosaccharide moieties of glycoproteins and these similarities will be pointed out in the following sections of the book when appropriate.

The three classes of complex carbohydrates, namely glycoproteins, proteoglycans and glycolipids, are often grouped together under the term *glycoconjugates* (Table 1.1).

In this book I have found it useful to adopt a shorthand notation to describe the complex carbohydrate sequences of glycans. Firstly, the monosaccharide constituents of glycoproteins to be discussed are always found in the pyranose form and, except for L-fucose, in the D-configuration (Fig. 1.2). Secondly, although the α- and β- forms are common in the glycans, there is a certain unity of structure and in some cases the anomeric configuration of monosaccharide constituents can be omitted without loss of clarity. Note the abbreviated forms of the monosaccharides used interchangeably throughout this book (Table 1.1).

2 Structure

2.1 Carbohydrate–protein linkages

Only a few side chains of the aminoacid residues usually found in proteins are suitable for the attachment of a mono- or oligo-saccharide (Table 2.1). A very common linkage involves the C(1) of *N*-acetylglucosamine and the amide group of asparagine. The structure (Fig. 2.1) is 2-acetamido-*N*-(L-aspart-4-oyl)-2-deoxy-*β*-D-glucopyranosylamine, trivial names 'asparaginyl-*N*-acetylglucosamine' or 'glcNAc.Asn'. The asparagine is part of an extended polypeptide sequence and the *N*-acetylglucosamine residue forms the point of attachment of other monosaccharides.

The carbohydrate chains attached to polypeptide in this fashion are *N*-glycans. In the second class of linkage sugars are joined through C(1) by glycosidic bonds to the hydroxylated side chains of serine, threonine, hydroxylysine or hydroxyproline. The latter aminoacids are commonly found in plants and L-arabinose linked glycosidically to hydroxyproline is the most common linkage found in plant glycoproteins [5]. D-Galactose may also be linked to hydroxyproline in some plant glycoproteins but no such linkage has been identified elsewhere. Hydroxylysine exists in vertebrate and invertebrate collagens and a unique linkage in which D-galactose is joined *β*-glycosidually to hydroxylysine is found. The disaccharide 2-*0*-*α*-D-glucopyranosyl D-galactose or glc $\alpha 1 \rightarrow 2$ gal is also found linked to the polypeptide

Table 2.1 Reactive aminoacid side chains

Aminoacid	*Functional group*	*Bond formed*	*Sugar linked*	*Presence in glycoproteins*	*Glycan*
Asparagine	Amide	Glycosylamine	glcNAc	Yes	*N*
Glutamic acid	Carboxyl	Ester		No	
Aspartic acid	Carboxyl	Ester		No	
Serine	Hydroxyl	Glycoside	galNAc, man, gal (Serine and Threonine)	Yes	*O*
Threonine	Hydroxyl	Glycoside		Yes	*O*
Hydroxylysine	Hydroxyl	Glycoside	gal	Yes	*O*
Hydroxyproline	Hydroxyl	Glycoside	ara, gal	Yes	*O*
Cysteine	Sulphydryl	Thioglycoside		No (?)	

ara, L-arabinose

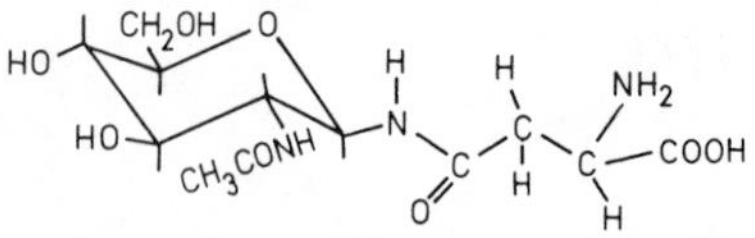

Fig. 2.1 Structure of the linkage of *N*-acetylglucosamine and asparagine. The hydrogen substituents of the sugar ring are omitted for convenience.

backbone in collagens (Fig. 2.2b). In the plant glycoprotein, extensin, galactose is attached to serine and a similar linkage is found in certain collagens for example of the primitive worms *Lumbricus* and *Nereis*. Mannose linked to serine and threonine forms the point of attachment of carbohydrate chains in glycoproteins synthesized by yeasts and moulds and has recently been detected in mammalian glycoproteins as minor components [6]. In the glycoproteins of higher organisms, however, the sugar most frequently found linked to serine and threonine is *N*-acetylgalactosamine (Fig. 2.2c). These carbohydrate chains are called *O*-glycans and the linkage is given the trivial name 'galNAc.ser(thr)-type' linkage.

2.2 Sequence analysis

Only rarely does a polypeptide moiety of a glycoprotein bear a single unique carbohydrate sequence. Usually several aminoacid residues are glycosylated and each glycosylated site may carry a mixture of closely similar glycans. Proteolysis will release these glycans attached to the linker aminoacid or to small peptides. The glycopeptides can be separated and analysed for monosaccharide composition and often their complete carbohydrate sequence is achieved [7] by a variety of chemical, enzymic or physical techniques (Table 2.2).

A special analytical problem concerns the *sialic acid* constituents of glycoproteins since the basic ring structure of neuraminic acid occurs in a great variety of forms differing in the positions of substituent groups. The most common derivative is *N*-acetylneuraminic acid (Fig. 2.3) but about twenty other derivatives are known. *N*-glycolylneuraminic acid,

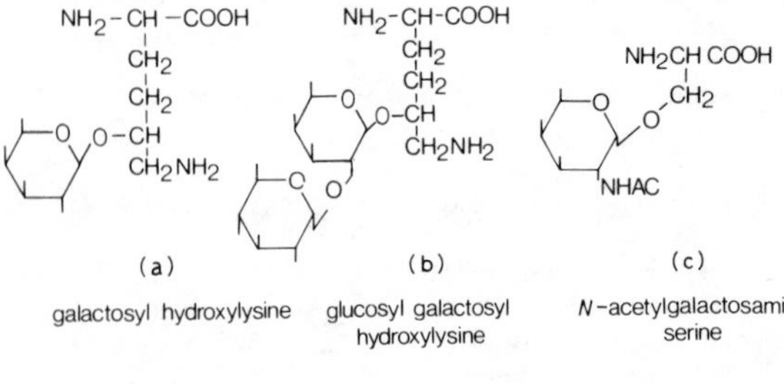

(a) galactosyl hydroxylysine

(b) glucosyl galactosyl hydroxylysine

(c) *N*-acetylgalactosaminyl serine

Fig. 2.2 Structure of linkages present in *O*-glycans. Only the configuration of non-hydrogen substituents on the sugar rings are indicated.

Table 2.2 Steps in the sequence analysis of glycans

1. Proteolysis: pronase, trypsin, chymotrypsin, collagenase, etc.
2. Separation of glycopeptides: gel filtration, ion-exchange chromatography, paper electrophoresis and chromatography.
3. Composition: gas–liquid chromatography (monosaccharides), ion-exchange chromatography (aminoacids).
4. Carbohydrate sequence determination:
   (a) chemical: methylation, periodate oxidation
   (b) enzymic: *exo*- and *endo*-glycosidases
   (c) physical: mass spectrometry, nuclear magnetic resonance.

for example, frequently occurs in glycoproteins or glycolipids and several *O*-acyl derivatives of this structure and *N*-acetylneuraminic acid exist. Glycoside formation involves the hydroxyl at C(2).

2.2.1 Enzymic analysis

Enzymes metabolizing complex carbohydrates are valuable tools with which to study structural features since these enzymes often display rather exact specificities for their substrates. So far the enzymes most widely used are the *glycosidases* [8], since many of these are available in purified form (Table 2.3). The list is not exhaustive and does not include the very important class of *endo*-glycosidases, discussion of which will be delayed because of the complexity of their substrate requirements.

All of the enzymes listed are *exo*-glycosidases: monosaccharides are cleaved from glycosidic linkages only when the sugars are terminal residues. Most *exo*-glycosidases display great specificity for the sugar moiety and the anomeric configuration of the glycosidic linkage. The example shown in Fig. 2.4 is an α-mannosidase and a β-mannoside would not be hydrolysed. In addition many *exo*-glycosidases hydrolyse at different rates the appropriate glycosides having different *aglycon* (R)

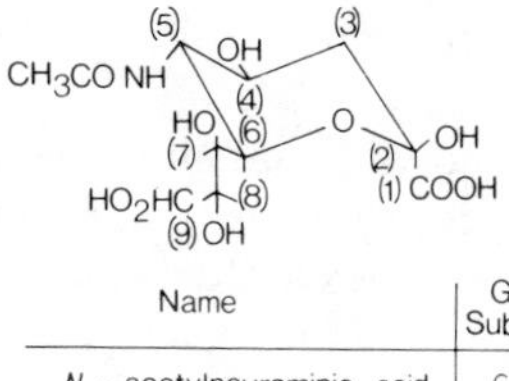

Name	Group Substituted	Substitution Group	Typical Source
N - acetylneuraminic acid	C (5)	CH_3CONH-	common
N - glycolylneuraminic acid	C (5)	HO CH_2CONH	pig tissues >90%
O - acetyl derivatives			
	C (9)	CH_3CO -	bovine submandibular gland (rare)
	C (7)	CH_3CO -	
	C (4)	CH_3CO -	horse tissue

Fig. 2.3 Structure of *N*-acetylneuraminic acid and other sialic acids.

Table 2.3 *Exo*-glycosidases used for determination of carbohydrate sequences in glycoproteins

Glycosidase	*Source*	*Linkage specificity*
Neuraminidase	*Streptococcus pneumoniae*	
	Clostridium perfringens	NeuNAc α2 → 3 (6) gal; NeuNAc α2 → 6 glcNAc
	Vibrio cholerae	NeuNAc α2 → 3 gal ≫ α2 → 6
	Influenza virus	NeuNAc α2 → 3 gal
	Arthrobacter ureafaciens	As *Clostridium perfringens*
β-Galactosidase	*Streptococcus pneumoniae*	Gal β1 → 4 glcNAc only
	Clostridium perfringens	
	Jack bean meal	Gal β1 → 4 glcNAc ≫ β1 → 3
	Aspergillus niger	
β-Mannosidase	Hen oviduct	Not defined
	Turbo cornutus	
	Snail	
	Pineapple	
α-Mannosidase	Jack bean meal	man α1 → 2 man = α1 → 6 ≫ α1 → 3
	Aspergillus niger	
α-Fucosidase	*Clostridium perfringens*	Fuc α1 → 2 gal only
	Aspergillus niger	
	Bacillus fulminans	Fuc α1 → 2 gal only
	Turbo cornutus }	All known fucosyl linkages
	Charania lampas }	
	Almond emulsin	Fuc α1 → 3(4) gal; not α1 → 2; not fuc α1 → 6 glcNAc
β-*N*-acetyl-glucosaminidase	*Streptococcus pneumoniae*	Broad
	Clostridium perfringens	
	Jack bean meal	

moieties (see Table 2.3 for well-defined examples). Fig. 2.5 illustrates a typical application of *exo*-glycosidases to determine the order of monosaccharide residues at the non-reducing termini of the carbohydrate chains of the major glycoprotein with neuraminidase orosomucoid. Treatment of the intact glycoprotein with neuraminidase from *Streptococcus pneumoniae* (formerly *Diplococcus pneumoniae*) liberates free sialic acid quantitatively showing that all of these residues are terminal. Treatment of the intact glycoprotein with *S. pneumoniae* β-galactosidase releases no free galactose. However after removal of sialic acid, β-galactosidase releases over 80% of the total galactose content of

CH$_2$OH — O-R ⟶ CH$_2$OH — OH + R-OH
AGLYCON
HYDROLYSIS

Fig. 2.4 Action of *exo*-glycosidases.

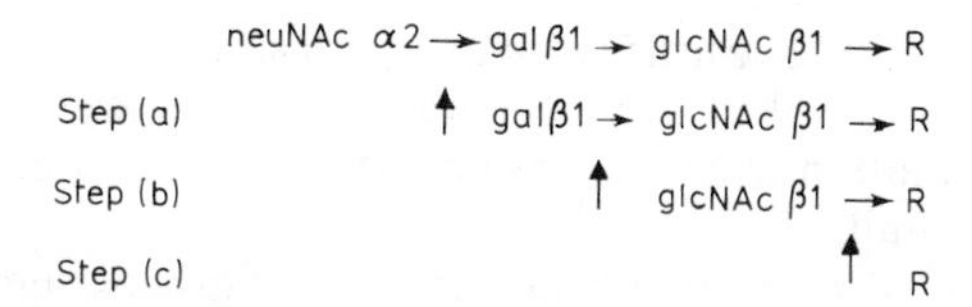

Fig. 2.5 Sequential degradation of orosomucoid with *exo*-glycosidases of *Streptococcus pneumoniae*. R = core oligosaccharide linked to the protein moiety.

the glycoprotein: the linkage must be $\beta 1 \rightarrow 4$ (Table 2.3). In the third step *N*-acetylglucosamine residues exposed by the *β*-galactosidase treatment in step (b) can be released by *β*-*N*-acetylglucosaminidase. The arrangement of monosaccharides is therefore determined by this simple experiment.

2.3 Structure of *N*-glycans

2.3.1 Ovalbumin

This major glycoprotein constituent of chick egg white has long been a favourite material for structural studies. Ovalbumin can be isolated in pure crystalline form, that is to say as material consisting of one particular polypeptide with one asparagine residue in a unique sequence to which carbohydrate is attached. Recently determination of the complete sequences of the carbohydrate chains of ovalbumin has had a powerful effect on formulating thinking about the biosynthetic assembly of *N*-glycans in general.

The structural analysis of the carbohydrate component of ovalbumin follows the procedures outlined previously (Table 2.2). Following exhaustive proteolysis with the non-specific protease, pronase, the glycopeptide fraction can be isolated by gel filtration. The aminoacids and small peptides produced by pronase are well separated from the carbohydrate–asparagine complex of larger molecular weight. Since ovalbumin contains a unique polypeptide that is substituted by one carbohydrate chain it might be expected that the carbohydrate fragment produced by pronase would be a pure compound or relatively simple mixture. However this was shown not to be the case: the glycopeptide fraction is separable into at least six components (I–VI) by ion-exchange chromatography [9]. This important observation leads to the conclusion that the carbohydrate unit attached to a particular aminoacid residue in a polypeptide may show structural heterogeneity. The result was surprising at the time and considerable effort was put into eliminating trivial reasons: (1) similar heterogeneity was detected using ovalbumin obtained from the eggs of a single chicken; (2) every genetic variant of chicken shared the same variety of glycopeptides; (3) the relative amounts of the glycopeptides did not depend on the age of the egg; (4) although chick egg white contains glycosidase activities, e.g. α-mannosidases and *β*-*N*-acetylglucosaminidases, the ovalbumins of birds such as turkey that have little or no glycosidase activities in the eggwhite also shared the several glycopeptide fractions. These cannot therefore be artefacts of isolation due to glycosidase degradation of a single original carbohydrate unit.

The six main glycopeptide fractions all contain asparagine as the only aminoacid and their monosaccharide compositions are rather similar in that all contain mannose and *N*-acetylglucosamine as the major (fractions I and II) or only (fractions III, IV, V and VI) constituents. Fractions II–IV are resolved further into two to three subfractions [10]. By a combination of physical, chemical and enzymic techniques the structures of each of the major ovalbumin glycopeptides have been assigned complete structures (Fig. 2.6). Strikingly all of the glycopeptides contain a common *core region* linked to asparagine. The glycopeptides VI, V, IV and III B form a series of closely related '*oligomannosidic*' or '*high mannose*' structures containing 4–7 mannose residues in addition to a *N*-acetylglucosaminyl $\beta 1 \rightarrow 4$ *N*-acetylglucosamine (chitobiose) unit linked to asparagine. In each case a mannose residue linked β-glycosidically to the chitobiose unit is itself substituted by two mannose units in α-glycosidic linkage to hydroxyls at C(3) and C(6) and these α-mannosyl residues form the points for further elongation of the chain by addition of mannose residues. The presence of terminal α-mannosyl residues in the glycopeptides can be readily demonstrated by treatment with α-mannosidases and determination of the amount of free mannose released. Almost exactly four and five residues per mole are released from glycopeptides V and IV respectively as predicted from the structures shown in Fig. 2.6. Hydrolysis in each case stops when the mannose residue linked β-glycosidically to *N*-acetylglucosamine is reached producing the same trisaccharide man$\beta 1 \rightarrow 4$ glcNAc $\beta 1 \rightarrow 4$ glcNAc linked to asparagine. By contrast virtually none of the *N*-acetylglucosamine residues are released as free monosaccharide by treatment with *exo*-β-*N*-acetylglucosaminidases. As explained previously the *exo*-glycosidases act only on terminal non-reducing residues and the *N*-acetylglucosamine units within the chitobiose core region are not susceptible to such enzymes. The results obtained with the *exo*-glycosidases are also fully consistent with the structures of the other ovalbumin glycopeptides. Several of the *N*-acetylglucosamine residues present in fractions I, II and III are released as the free sugar by *exo*-β-*N*-acetylglucosaminidase while fewer of the mannose residues are susceptible to α-mannosidase. In these '*hybrid*' *chains* the terminal β-*N*-acetylglucosamine residues occupy two major positions: (1) substituted $\beta 1 \rightarrow 4$ on to the mannose residue of the core region and (2) substituted $\beta 1 \rightarrow 2$ or $\beta 1 \rightarrow 2$ *and* $\beta 1 \rightarrow 4$ on to α-mannosyl residues contiguous to the core region. In the second case some of the *N*-acetylglucosamine units are substituted further by galactose to produce the *N*-acetyllactosamine sequence; gal $\beta 1 \rightarrow 4$ glcNAc first detected in orosomucoid (otherwise called α_1,-acid glycoprotein).

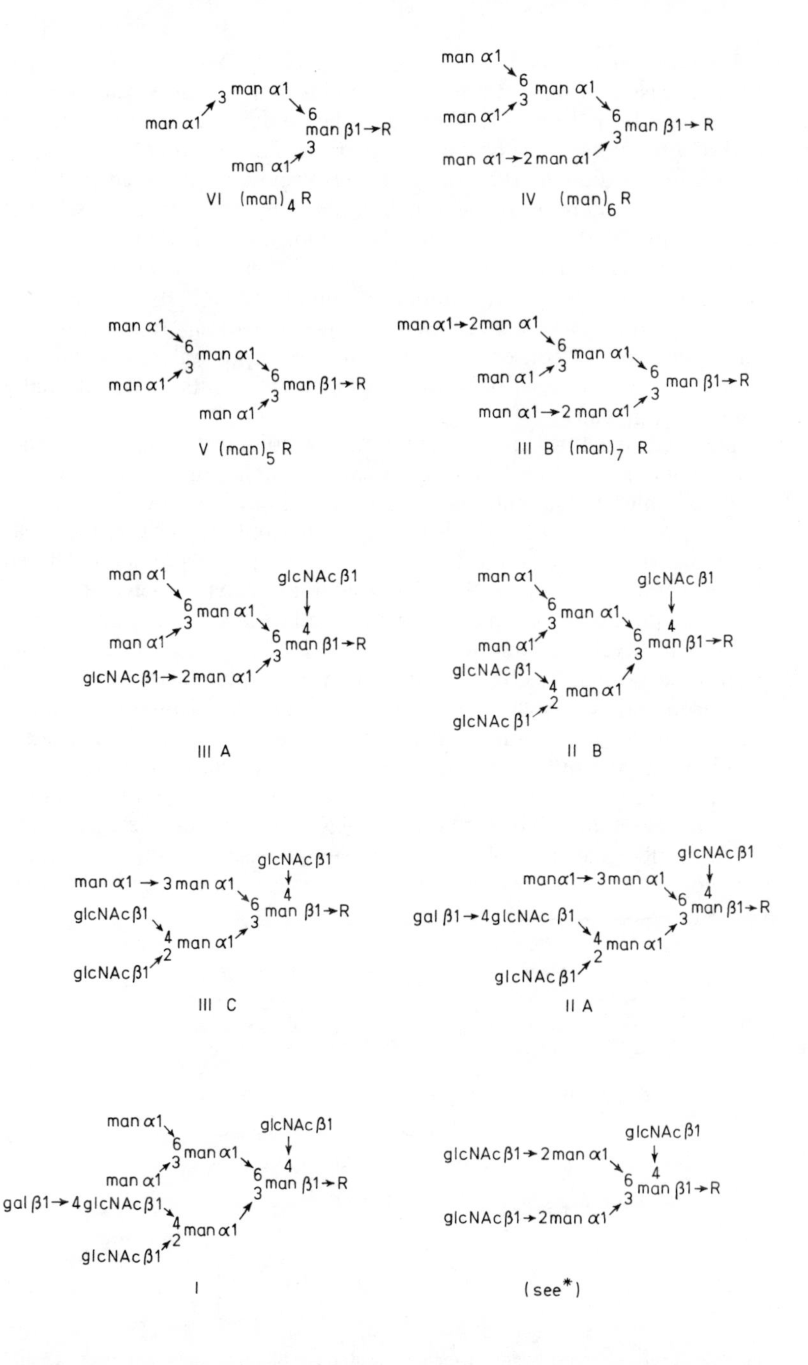

Fig. 2.6 Structures of ovalbumin glycopeptides. R = glcNAc β1 →4 glcNAc . Asn. * A minor component present in fraction V and isolated by electrophoresis [10].

2.3.2 Immunoglobulins

All five classes of human immunoglobulins and probably other species contain carbohydrate [11]. Although the content and composition of carbohydrate may vary between the various classes, all appear to carry *N*-glycans located in the constant (Fc) region of the heavy chains. Interestingly there is considerable homology in the exact sites of substitution of the Fc piece with the carbohydrate units (Fig. 2.7).

The fact that homologous carbohydrate chains are located at similar points in the Fc region of all immunoglobulins strongly suggests that the necessary aminoacid sequences surrounding the asparagine residues have been highly conserved and points to important structural and functional significance, possibly as attachment points for functionally significant glycans.

Detailed structural analysis of the two carbohydrate chains of human immunoglobulin M and the single chain of human immunoglobulin G are available. These analyses were done using immunoglobulins prepared from the sera of two myeloma patients and may be considered as referring to the secreted products of cloned cells, that is to say of defined genes in each case. Nevertheless considerable heterogeneity of carbohydrate structure is present exactly as found for the ovalbumin glycans.

Human myeloma immunoglobulin M contains at each asparagine residue carbohydrate chains of two main types: chains containing only mannose and *N*-acetylglucosamine and chains containing in addition sialic acid, galactose and fucose. The structures of the latter are not available as yet but the structures of the oligomannosidic chains are shown in Fig. 2.8. Two major points are to be made. Firstly all of the sequences are built up by the addition of α-mannosyl residues to the core trisaccharide man $\beta 1 \rightarrow 4$ glcNAc $\beta 1 \rightarrow 4$ glcNAc; first by the attachment of one mannose residue to C(3) and three mannose residues to C(6)

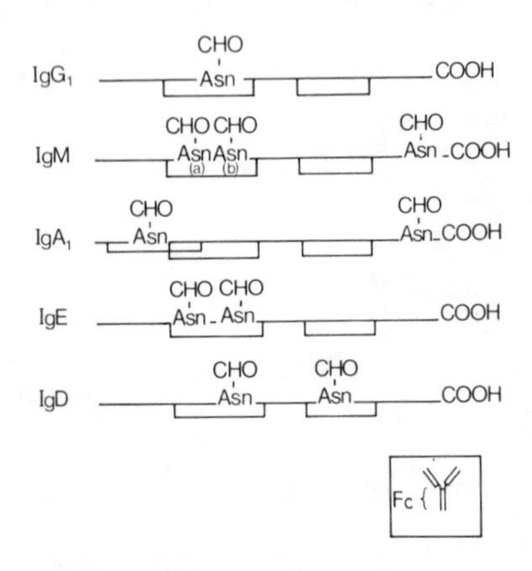

Fig. 2.7 Glycosylation sites and intrachain disulphides of immunoglobulin heavy chains. The *inset* shows the intact immunoglobulin molecule. CHO, carbohydrate units. In IgM the two units (a) and (b) are discussed in the text.

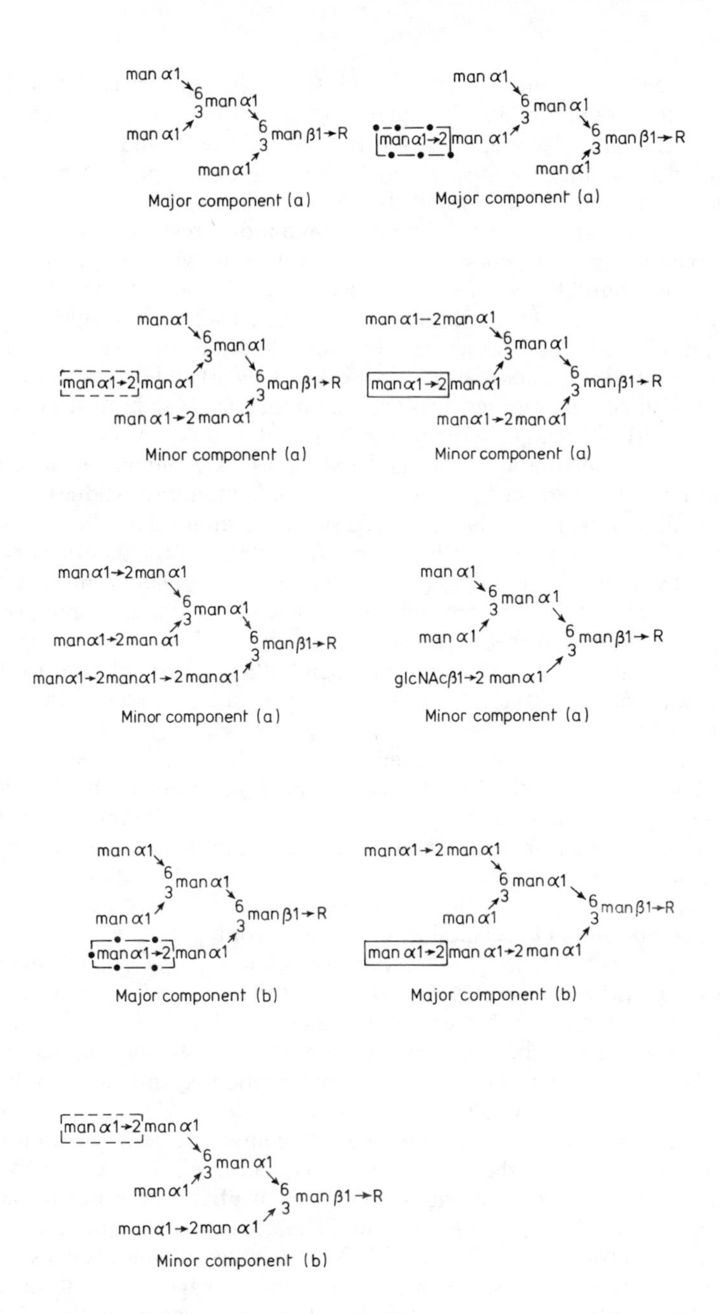

Fig. 2.8 Structures of the *N*-glycans of human immunoglobulin M. R = glcNAc $\beta 1 \rightarrow 4$ glcNAc. Asn; (a) and (b) are the two glycosylated sites on the Fc portion as shown in Fig. 2.7. Variable α-mannosyl residues are indicated by the paired boxes. (From Chapman and Kornfeld [13].)

of the central β-mannosyl residue. This $(man)_5R$ sequence is identical in structure to ovalbumin glycopeptide V (Fig. 2.6) and structures identical to ovalbumin glycopeptides III B and IV are found among the immunoglobulin carbohydrate chains (the minor component (b) and a major component (b) respectively.)

Larger structures containing more α-mannosyl residues than any of the ovalbumin glycopeptides are present however; two $(man)_8R$ isomers and one $(man)_9R$ sequence. The second very important point is the asymmetry of the glycosylation at sites (a) and (b) of the IgM heavy chain. This is best described by consideration of the structures containing six mannose residues present at sites (a) and (b) respectively. Five of these mannose residues are placed identically in both structures but one $\alpha 1 \rightarrow 2$ mannosyl unit is substituted at different points in the carbohydrate chains at the two glycosylation sites. A similar asymmetry is seen for structures containing seven and eight mannose residues which again differ only in the placement of a single α-mannosyl residue. We are far from understanding the reason for these subtle differences in carbohydrate structure of glycans of similar gross structure present on asparagine residues spaced quite closely together on the same polypeptide chain. However, the structures obtained for the oligomannosidic chains of this human immunoglobulin are very relevant to the pathways of biosynthesis of *N*-glycans and will be discussed in the next chapter.

Finally mention should be made of the minor component at site (a) containing a terminal *N*-acetyl-glucosamine residue. As will be seen this is an important intermediate in the assembly of the 'complex' glycans containing galactose, fucose and sialic acid present in immunoglobulins and other glycoproteins.

Analysis of glycopeptides purified from pronase digests of an immunoglobulin G prepared from the serum of a patient with multiple myeloma has produced the structures shown in Fig. 2.9 [10]. Chromatography of the glycopeptides on Dowex 50 gives an acidic glycopeptide (a) that passes through the column quickly and a mixture of neutral glycopeptides (e.g. b, c) that can be separated for structural analysis. Sequential treatment of all of these glycopeptides with neuraminidase (for (a) only), β-galactosidase and β-*N*-acetylglucosaminidase produces the same oligosaccharide (d) containing two α-mannosyl residues linked to the core region sequence. Clearly the heavily mannosylated oligosaccharides present in ovalbumin are not present in major amounts in immunoglobulin G. The two α-mannosyl residues are substituted however with *N*-acetylglucosamine residues and galactose residues, in some cases in the same linkages as found in the ovalbumin glycopeptides, e.g. fraction I. An interesting feature of the immunoglobulin glycopeptide (a) is its asymmetry. The single *N*-acetylneuraminic acid residue is present on the side-chain linked by $\alpha 1 \rightarrow 3$ linkage to the central β-linked mannose unit of the core region. This structure raises the question of specificity of the glycosyl transferase

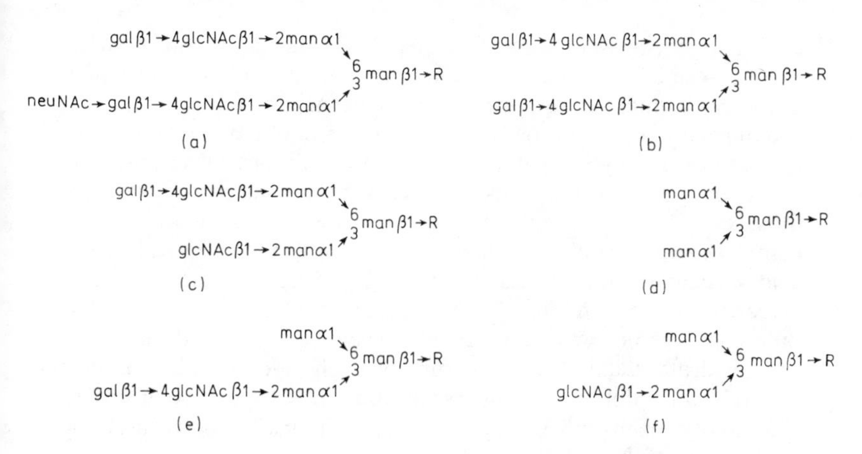

Fig. 2.9 Structures of human myeloma IgG glycopeptides and products of *exo*-glycosidase degradation. R = glcNAc $\beta 1 \rightarrow 4$ (fuc $\alpha 1 \rightarrow 6$) glcNAc . Asn.

attaching sialic acid residues to terminal galactose of such carbohydrate chains as discussed in the next chapter. The occurrence of the asymmetric structure is of practical value in providing substrates for specific glycosyl transferases. Degradation of the acid glycopeptide (a) with β-galactosidase followed by β-*N*-acetylglucosaminidase and then neuraminidase produces structure (e) and further degradation with β-galactosidase produces structure (f). These structures do not occur naturally in immunoglobulin but a structure analogous to (f) is present in immunoglobulin M as previously described. An additional new feature of these *N*-glycans is the fucose substituent on the chitobiose unit of the core sequence; it is linked $\alpha 1 \rightarrow 6$ to the first *N*-acetylglucosamine residue of the chain. It is important to note that the glycopeptides described refer to the IgG secreted by the particular myeloma studied [10, 11]. Each myeloma is different and produces rather different spectra of glycan structures.

2.3.3 Orosomucoid

Some structural features of this major serum constituent have already been discussed. The single polypeptide chain contains five glycosylated asparagine residues arranged as shown in Fig. 2.10. Recently the

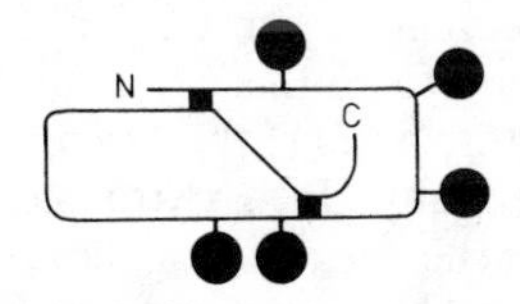

Fig. 2.10 Glycosylation sites (closed circles) on the single polypeptide moiety of orosomucoid. The two intra-chain disulphide bonds are indicated by the *thick* lines.

complete sequences of the desialylated *N*-glycans of orosomucoid have been described and placed in the linear aminoacid sequence of the polypeptide chain of the molecule. Sialic acid is removed from the glycoprotein before preparation of glycopeptides by treatment with chymotrypsin. Since chymotrypsin has a rather narrow specificity this procedure gives rise to glycopeptides each containing one of five peptide moieties with unique sequence that can be readily identified in the complete sequence of the intact polypeptide chain. Removal of sialic acid was carried out in this case to make easier separation of sixteen glycopeptides on an ion-exchange column taking advantage of the differences in net charge of their peptide moieties and size of the carbohydrate chains. The structure of each glycopeptide was determined by interpretation of high resolution NMR spectra using a total of 2–3 mg of each glycopeptide: a remarkable indication of the usefulness of this powerful analytical tool [14].

The various structures show several new features as well as confirming the conclusions drawn from degradation of the intact glycoprotein as described in Section 2.2.1. The sialic acid residues presumably are substituted onto the *N*-acetyllactosamine sequences present at the terminus of the branched oligosaccharides. No oligomannosidic sequences are found in orosomucoid and the chains are totally of the 'complex' type containing multiple copies of the neuNAc . gal . glcNAc sequence substituted onto the common central region consisting of two α-mannosyl units, one β-mannosyl unit and a chitobiose unit. Again as with the 'complex' chain of immunoglobulin G there is a clear asymmetry in the structures containing three or more of the terminating *N*-acetyllactosamine units. The α-mannosyl unit linked to C(3) of the β-mannose residue is substituted at C(2) and C(4) whereas the other α-mannosyl residue is substituted on C(2) and C(6). Such asymmetry most likely arises from the substrate specificities of separate *N*-acetylglucosaminyl transferases responsible for attachment of each *N*-acetylglucosaminyl unit in the precursor sequence during chain assembly. The substrate specificities of two of these enzymes are known in some detail as discussed in Chapter 3.

Another novel feature of the orosomucoid *N*-glycans is the presence of fucose substituted $\alpha 1 \rightarrow 3$ to a unique *N*-acetylglucosamine unit. The fact that only particular *N*-acetylglucosamine units in the outer branches are substituted again points to an exact specificity of the $\alpha 1 \rightarrow 3$ fucosyl transferase. Notice that fucose is not attached to the chitobiose sequence, a structure present in immunoglobulin G.

The conclusions arising from this brief summary of existing knowledge of *N*-glycan structure are: (1) the assembly of *N*-glycans involves the concerted action of many glycosyl transferases each of very exact substrate specificity; (2) the activities of these enzymes are modulated significantly depending on the exact site in a polypeptide chain being glycosylated. Some biosynthetic evidence for these conclusions is given in the next chapter.

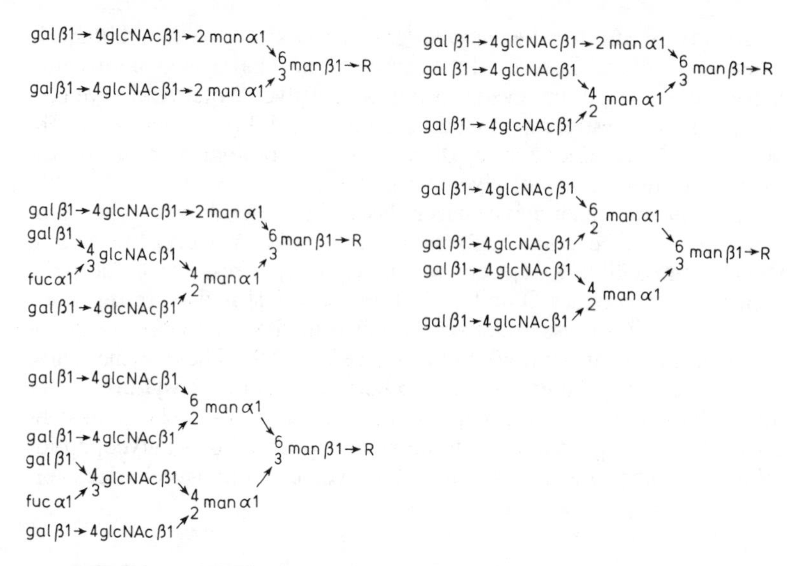

Fig. 2.11 Structure of orosomucoid *N*-glycans. R = glcNAc β1 → 4 glcNAc . Asn.

2.4 *Endo*-glycosidases

As discussed briefly in Section 2.2.1 *endo*-glycosidases hydrolyse internal linkages and release oligosaccharides from glycoproteins or glycopeptides obtained from glycoproteins by proteolysis. Many *endo*-glycosidases have been identified (Table 2.4) [8]. Characterization of the specificity of the enzymes has relied on the availability of oligosaccharides or glycopeptides of known structure such as the ovalbumin

Table 2.4 *Endo*-glycosidases [8]

Enzyme	*Source*
Endo-β-N-acetylglucosaminidase Type A	
D	*Streptococcus pneumoniae*
H	*Streptomyces plicatus*
	Streptomyces griseus
CI and CII	*Clostridium perfringens*
	Fig
	Mammalian kidney, liver, spleen
	Hen oviduct
Type B	Almond emulsin
Endo-β-N-acetylgalactosaminidase	
	Streptococcus pneumoniae
	Clostridium perfringens
Endo-β-galactosidase	*Streptococcus pneumoniae*
	Escherichia freundii

fragments. In turn the specific *endo*-glycosidases have been invaluable in defining structural features of the carbohydrate chains of glycoproteins. Since some of these enzymes react with intact glycoproteins and even the glycoproteins present on intact cells, their availability has opened up the possibility of assessing in a direct way the biological roles of the carbohydrate moieties of glycoconjugates.

The *endo-β-N*-acetylglucosaminidases Type A all split the same glycosidic linkage (Fig. 2.12) between the two *N*-acetylglucosamine residues immediately adjacent to the asparagine residue of the polypeptide. The enzymes D and CI but not H or CII hydrolyse this bond even if the chitobiose sequence is substituted with fucose as in the glycopeptides of immunoglobulin G (see Fig. 2.9). The enzymes show further specificity differences. The α3-substituted mannosyl unit at the non-reducing end of glycopeptides susceptible to D and CI must be unsubstituted, e.g. as in ovalbumin glycopeptide VI or V; glycopeptide IV is not a substrate. The enzymes also hydrolyse oligosaccharides, e.g.

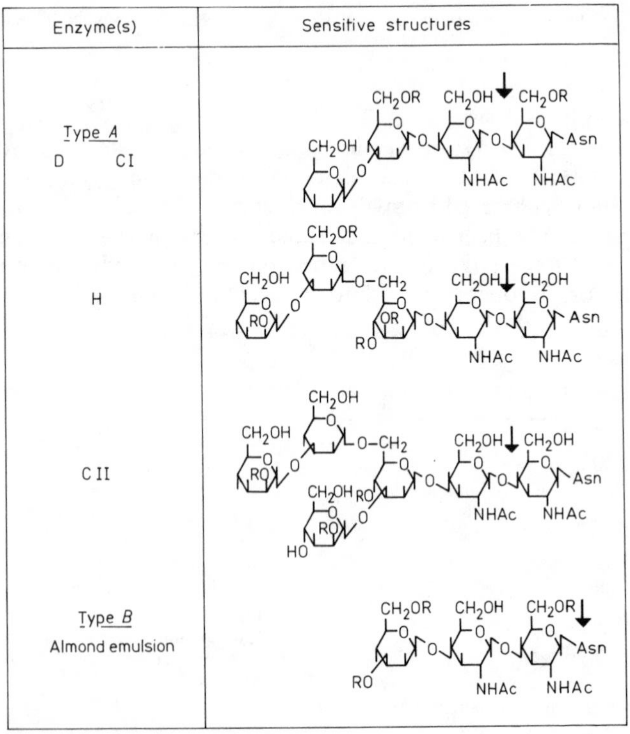

Enzyme(s)	Sensitive structures
Type A D CI	CH_2OR CH_2OH CH_2OR CH_2OH O O O O Asn NHAc NHAc
H	CH_2OR CH_2OH O–CH_2 CH_2OH CH_2OH RO O OR RO O O Asn NHAc NHAc
C II	CH_2OH CH_2OH O–CH_2 CH_2OH CH_2OH RO O CH_2OH RO RO O HO O O Asn NHAc NHAc
Type B Almond emulsion	CH_2OR CH_2OH CH_2OR O O O Asn RO NHAc NHAc

Fig. 2.12 Specificity of *endo-β-N*-acetylglucosaminidases. R represents either hydrogen or sugar residue.

products of hydrolysis by type B *endo*-β-*N*-acetylglucosaminidase. Glycosidase H requires a more complex sequence of α-mannosyl residues substituted onto the C(6) of the β-mannosyl residue of the trisaccharide core of glycopeptides and does not react when fucose is present in the core region. Ovalbumin glycopeptides IIIA, IIIB, IV, V and VI are rapidly cleaved by this enzyme and by glycosidase CII. There are two major differences between enzymes H and CII. The latter enzyme requires a branched structure in which both C(3) and C(6) of the β-mannosyl unit are substituted with α-mannosyl residues. Secondly CII enzyme reacts poorly with structures containing a 4-*O*-substituted mannose residue linked α3 to the core sequence: position 4 must carry a free hydroxyl as shown in Fig. 2.12 [11].

The type B *endo*-β-*N*-acetylglucosaminidase, e.g. of almond emulsin [15] is reactive with both intact glycoproteins and glycopeptides. The bond split is that between *N*-acetylglucosamine and asparagine (Fig. 2.12).

The *Streptococcus* endo-β-*N*-acetylgalactosaminidase is highly specific for the sequence gal β1 → 3 galNAc linked to serine or threonine residues of polypeptides, the basic feature of many *O*-glycans of glycoproteins. Substitution of the disaccharide in any way completely abolishes activity thereby limiting the usefulness of the enzyme. A somewhat similar enzyme is obtained from culture fluids of *Clostridium perfringens*. The products of the reaction are the free gal β1 → 3 galNAc disaccharide and the polypeptide chain.

The *endo*-β-galactosidases are also specific for extended carbohydrate sequences. The *E. freundii* enzyme hydrolyses an internal galactosidic bond in carbohydrate chains of glycoproteins when galactose is linked β1 → 4 to *N*-acetylglucosamine or in glycolipids when the linkage is β1 → 4 to glucose which in turn is joined to a ceramide unit. The internal linkage is cleaved even in very long oligosaccharide chains formed by addition of sugar residues to the non-reducing terminal *N*-acetylglucosamine unit of the structure shown in Fig. 2.13. Substitution

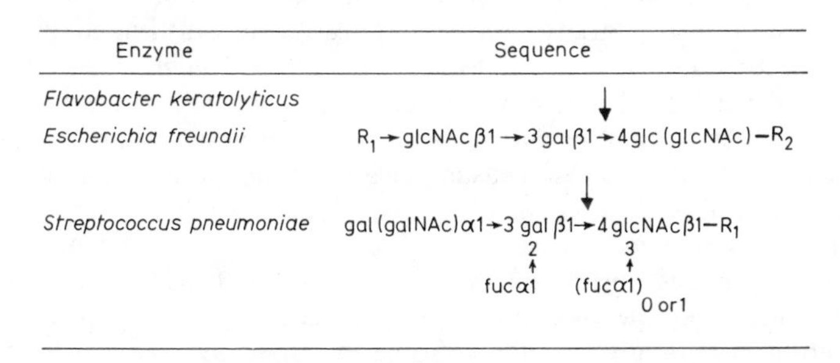

Fig. 2.13 Substrate specificities of *endo*-β-galactosidases. R_1 = hydrogen or sugar; R_2 = hydrogen, sugars or ceramide.

at C(6) of the galactose residue in the essential trisaccharide sequence completely abolishes susceptibility to the enzyme. The *N*-acetylglucosamine residue may be sulphated however and the structure remains susceptible. This particular sulphated structure is present in *keratansulphate*, a constituent of proteoglycans, which is hydrolysed by the enzyme to release free chains of neuNAc → gal → glcNAc (SO_4) → gal. The *Flavobacter* enzyme has a specificity similar to that just discussed and both enzymes have been purified to homogeneity, being proteins of molecular weight 35 000. One practical difference between these enzymes is that in *Flavobacter* the enzyme is constitutive, while synthesis and secretion in *Escherichia freundii* is only induced by growth of the organism on keratansulphate or glycoproteins with the appropriate linkages.

The *Streptococcal endo-β*-galactosidase hydrolyses the branched oligosaccharides shown in Fig. 2.13. These structures represent the antigenic determinants of human blood groups A (with *N*-acetylgalactosamine terminal unit) or B (with galactose terminal unit). The *N*-acetylglucosamine is an essential part of the susceptible structures and only gal β1 → 4 glcNAc linkages but not gal β1 → 3 glcNAc linkages are hydrolysed. Both of these linkages are present in separate chains substituted with fucose and either galactose or *N*-acetylgalactosamine to form distinct chains with A and B blood group activity. It is interesting therefore that the *Streptococcus* enzyme can discriminate antigenic structures and exhibits higher specificity than antibodies in this case.

2.5 Structure of *O*-glycans

In addition to the *O*-glycans of collagen discussed previously several structural features of glycans linked glycosidically to hydroxy-aminoacids are known.

2.5.1 Glycoprotein secretions (mucins)

Anyone who has seen the slime left by a mobile snail on a dry road or has eaten 'bird nest soup' in Chinese restaurants is familiar with glycoproteins containing large quantities of *O*-(glycosidically linked) glycans. These and similar glycoproteins are often called *mucins* or *mucus* (the German chemist Hoppe-Seyler coined the even less complimentary name 'Schleimstoff'), because of their extreme viscosity and jelly-like appearance. Although unpleasant in appearance and to handle, and extremely intractable to chemical and physical analysis, the 'mucins' nevertheless play vital roles. It seems to be of great importance in all higher organisms, vertebrate and invertebrate, that the body cavities of the respiratory, digestive and urogenital tracts are lined by a cell layer secreting viscous glycoproteins. Some examples are saliva (the basic ingredient of 'bird nest soup'), bronchial and intestinal secretions, seminal plasma and cervical mucus. In addition to roles as lubricants, for instance in the digestive tracts, a thick viscous layer affords protection to the underlying tissue against mechanical or chemical insult, e.g. gastric juices.

In view of their biological importance and association with severe disease states, e.g. cystic fibrosis, the glycoprotein secretions have been studied intensively for many years. The salivary gland secretions are best characterized from the point of view of carbohydrate composition, particularly the glycoprotein secretions of the submaxillary gland. The simplest carbohydrate structure is found in sheep glands, studied thoroughly in Australia by Alfred Gottschalk twenty years ago and later in Germany. The disaccharide consisting of *N*-acetylneuraminic acid and *N*-acetylgalactosamine is linked to many serine and threonine residues in the polypeptide moiety of the glycoprotein (Fig. 2.14). About 200 disaccharides are scattered along a single polypeptide chain; one out of every six aminoacid residues being a glycosylated serine or threonine. Proline is the third most common aminoacid residue in the submaxillary glycoprotein and the sites of glycosylation are therefore very rich in proline. The glycoprotein is a comb-like structure with short carbohydrate side chains for teeth on a rigid (proline-rich) polypeptide handle. These comb-like structures, combined together by disulphide bridges between their protein cores, give larger glycoprotein molecules with the peculiar viscous properties [4]. The polypeptide moiety of submaxillary glycoproteins of species other than sheep carry large numbers of oligosaccharide side chains in addition to the disaccharide just discussed (Fig. 2.14). In pig submaxillary gland secretions which have been particularly well characterized a neutral disaccharide gal $\beta 1 \rightarrow 3$ galNAc is present and sialylated forms are also found. Of special significance is

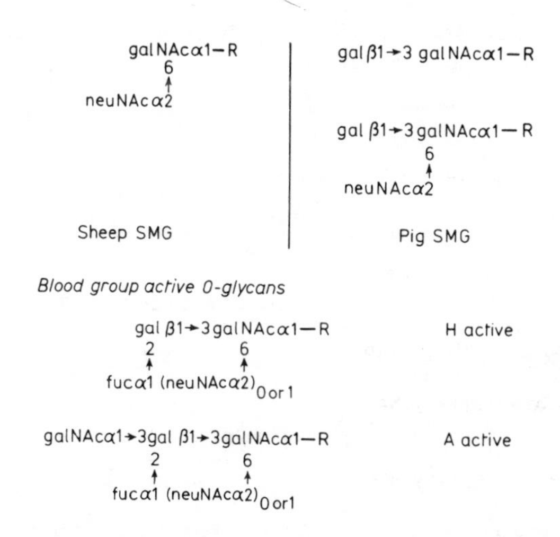

Fig. 2.14 Structure of *O*-glycans containing the *N*-acetyl-galactosaminyl serine (threonine) linkage in submaxillary gland (SMG) secretions. R = serine or threonine residue in a polypeptide.

the presence in the porcine glycoprotein of chains containing fucose. These chains are easily recognized since they occur also in human secretions as a human blood group antigen H. Substitution of this sequence with an *N*-acetylgalactosamine residue converts the H antigenic determinant into a sequence known as blood group A and this sequence too is present in pig submaxillary gland glycoproteins (Fig. 2.14).

Although mucus secretions of the stomach and intestine are readily available, very little is known because these secretions are a mixture of many glycoproteins which are difficult to separate. Some recent structures characterized [16] in glycoproteins from rat colon are shown in Fig. 2.15 which illustrate new features of *O*-glycans. The key new linkage is the *N*-acetylglucosaminyl $\beta 1 \rightarrow 3$ *N*-acetylgalactosamine sequence. The *N*-acetylglucosamine residue forms the site of elongation into very complex sequences which in the largest chains are terminated by the sequence characteristic of the A-determinant (compare with the A-active sequence present in porcine submaxillary gland secretions, Fig. 2.14).

2.5.2 Human blood group substances

Strictly the term 'blood groups' refers to antigenic differences detected on the erythrocytes of different individuals by means of specific antibodies [2]. The major ABO(H) blood groups of man were discovered early in the century by Landsteiner. They forced attention because of naturally occurring antibodies, the origin of which is still obscure. Thus, people with the A blood group antigen on their red cells carry a high

```
glcNAcβ1→3galNAc→R                    galβ1→4glcNAcβ1→3galNAc→R
          6                                                6
          ↑                                                ↑
     neuNAcα2                                        (neuNAcα2)
                                                         0-1

Blood group active glycoproteins

galNAcα1→3galβ1→3glcNAcβ1→3galβ1→4glcNAcβ1→3galNAc→R       A active
          2                    6                   6
          ↑                    ↑                   ↑
       fucα1              neuNAcα2            neuNAcα2

galNAcα1→3galβ1→3glcNAcβ1
          2     (4)        ↘6
          ↑                  galβ1→4glcNAcβ1→3galNAc→R          A active
       fucα1               ↗3                     6
neuNAcα2→6galβ1→4glcNAcβ1                         ↑
                                             neuNAcα2

galNAcα1→3galβ1→3glcNAcβ1
          2     (4)        ↘6
          ↑                  galβ1→4glcNAcβ1→3galNAc→R          A active
       fucα1               ↗3                     6
galNAcα1→3galβ1→4glcNAcβ1                         ↑
          6                                  neuNAcα2
          ↑
     neuNAcα2
```

Fig. 2.15 Structures of *O*-glycans in rat colonic glycoproteins. R = serine or threonine residues in polypeptide. (From [16].)

concentration of antibodies against the B determinant and *vice versa*. People of the O group carry antibodies against both *A* and *B* determinants. Obviously, this phenomenon has had an enormous impact on the development of blood transfusion practice; the catastrophic consequence of mismatching blood group types are obvious. As a result of the extensive work of Walter Morgan, Winifred Watkins, Elvin Kabat and many others we know in detail the structural basis for ABO(H) antigenicity which resides in carbohydrate structure and ultimately in the genetic control of assembly of these structures. It is now known that glycoproteins present in many body sites also carry antigenic determinants related to the red cell antigens so the biological functions of the carbohydrate sequences and the meaning of their strictly-preserved inheritance in man, whatever these functions turn out to be, are not confined to expression only on the erythrocyte surface.

The product secreted by persons of *O*-genotype is called H-substance. The structure (Fig. 2.16) includes as an essential feature a fucose residue attached to C(2) of a galactose residue. The galactose is joined to either C(3) *or* C(4) of the next monosaccharide *N*-acetylglucosamine, in the sequence. Certain rare individuals first identified in India and hence the condition is called the Bombay phenotype, lack this fucose substituent and hence any A, B or H antigenicity on the red cells or in secreted glycoproteins (Fig. 2.16).

The H-active determinant may be converted into *either* an A- or B-active determinant by attachment respectively of an *N*-acetylgalactosamine or galactose residue in $\alpha 1 \rightarrow 3$ glycosidic linkage.

It is very clear that the A, B and H determinants are attached to the termini of *O*-glycans in the secreted glycoprotein. What then is the nature of the core sequence of these *O*-glycans? We have already seen

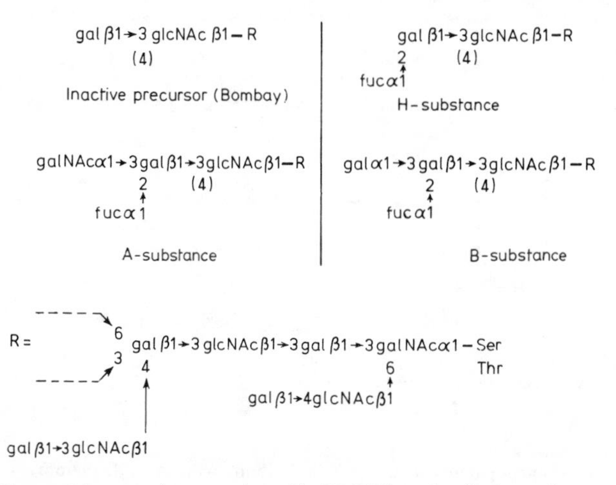

Fig. 2.16 Secreted human glycoproteins with ABO(H) antigenic determinants.

that many possibilities exist. In the pig submaxillary gland secretions the antigenic sequences are attached directly to *N*-acetylgalactosamine linked to polypeptide (Fig. 2.14). In the glycoproteins of rat colon (Fig. 2.15) a much longer core sequence is terminated by the antigenic determinants. In human secretions the core region of A, B and H active glycoproteins has not been fully worked out but a suggested structure for the products secreted in ovarian cysts is shown in Fig. 2.16.

It must be emphasized that these structures refer to *secreted* glycoproteins. What then are the blood group active substances of the erythrocyte membrane, originally defined by Landsteiner. Although still an unresolved problem it seems that the majority of the antigenic sequences are not present in glycoproteins at all but in glycolipids [17]. The glycolipids are based on the structures shown in Fig. 2.17: only the gal $\beta1 \rightarrow 4$ glcNAc type chains are found. Structures (a)–(d) are the simplest blood group active glycolipids but other more complex structures are also present, e.g. (e) and (f).

The essential feature of these structures is a repeat sequence of *N*-acetylglucosaminyl $\beta1 \rightarrow 3$ galactose linked through $\beta1$–4 linkages. The A, B and H structures are elaborated by attachment of the usual monosaccharides, fucose and *N*-acetylgalactosamine or galactose to a terminal galactose linked $\beta1 \rightarrow 4$ to the repeating sequence. When these substituents are missing another human blood group system is revealed

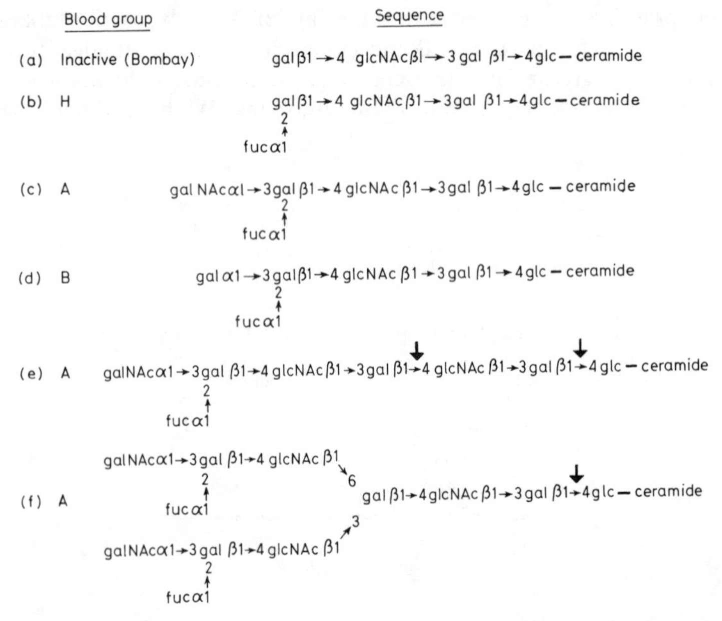

Fig. 2.17 Blood group active glycolipids of the human erythrocyte membrane. The *thick* arrows indicate bonds sensitive to *E. freundii endo-β* galactosidase.

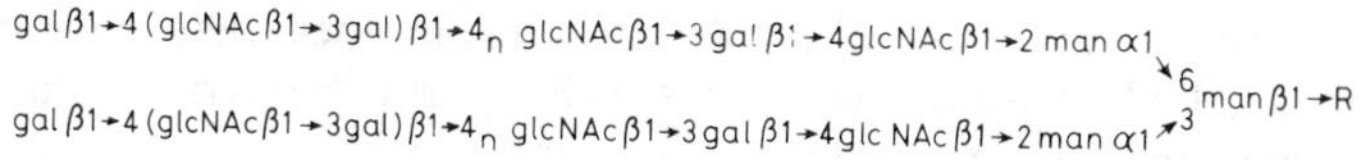

gal β1→4 (glcNAc β1→3gal) β1→4$_n$ glcNAc β1→3 gal β1→4glcNAc β1→2 man α1
↘6 man β1→R
gal β1→4 (glcNAc β1→3gal) β1→4$_n$ glcNAc β1→3 gal β1→4glc NAc β1→2 man α1 ↗3

i antigen

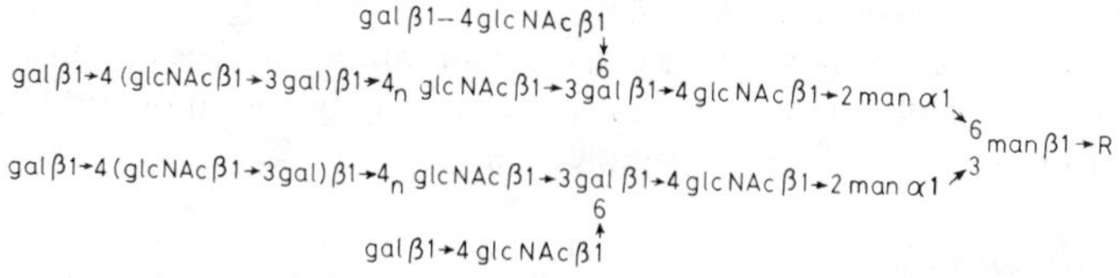

gal β1–4glc NAc β1
↓
6
gal β1→4 (glcNAc β1→3gal) β1→4$_n$ glc NAc β1→3gal β1→4 glc NAc β1→2 man α1
↘6 man β1→R
gal β1→4 (glcNAc β1→3gal) β1→4$_n$ glcNAc β1→3gal β1→4 glc NAc β1→2 man α1 ↗3
6
↑
gal β1→4 glc NAc β1

I antigen

Fig. 2.18 Structure of 'polyglycosyl' *N*-glycans of glycoproteins. R = glcNAc β1 → 4 glcNAc . Asn.

known as the Ii antigens. As in the original investigations of Landsteiner defining the ABO(H) human blood groups the Ii antigens are defined by antibodies occurring naturally in the blood of certain individuals. It now seems probable that the I and i specific antibodies recognize respectively the branched (f) and straight (e) – chain forms of the sequences.

Evidence is increasing that large amounts of blood group active determinants occur also on the glycoproteins of the erythrocyte membrane and other cell surface membranes. The antigenic determinants are carried on extended sequences similar to those present in the complex glycolipids (e) and (f). These sequences are susceptible to *E. freundii endo-β*-galactosidase, and when erythrocytes are treated with this enzyme the cells loose completely Ii antigenicity and a glycoprotein known as 'Band 3' [see Section 2.7] is converted into a carbohydrate-poor component of lower molecular weight. Other evidence shows that there is one such carbohydrate chain present in this glycoprotein [17]. These extended carbohydrate sequences are known as '*polyglycosyl*' units and have been identified as constituents of membrane-bound glycoproteins on several cell types, for example fibroblasts, as very minor components in addition to their presence in the erythrocyte Band 3 glycoprotein. The polyglycosyl units in glycoproteins are attached to a core region of *N*-glycan as shown in Fig. 2.18. In the I-active polyglycosyl chains several galactose residues are branched with the basic repeating sequence. These branched polyglycosyl chains predominate in adult erythrocytes while fetal cells contain the straight chain forms based on i antigen. Evidently the branching enzyme is under developmental control but we are far from understanding the significance of this for normal erythroid cell maturation.

2.6 Membrane glycoproteins: an integrated view

So far we have discussed the structure of glycoproteins obtained from fluids such as blood or in secretions from cells. Glycoproteins are also important constituents of all cellular membranes that have been examined [18]. Plasma membranes are a particularly rich source of glycoproteins and much of the current emphasis in glycoprotein research is on the structure of these glycoproteins, their organization within the membrane and the function of membrane-bound glycoproteins, in particular the carbohydrate moieties. The human erythrocyte, because it contains just one membrane, has been intensively studied and yielded much information that may be extrapolated to the membrane systems of more complex cells.

2.6.1 Glycophorin

The human red cell membrane consists of relatively few major proteins. Of these, three are particularly rich in carbohydrate and can be isolated by preferential extraction for example with chloroform–methanol mixtures, phenol or lithium diiodosalicylate followed by fractionation on the basis of molecular weight [19]. These glycoproteins are called glycophorin A, B and C; form A represents the major glycoprotein constituent. The complete structure of glycophorin A is shown in Fig. 2.19.

The glycoprotein subunit of molecular weight 29 000 contains a single polypeptide chain of 131 aminoacids. There are two variations in the aminoacid sequence at residues 1 and 5 as shown. These have biological significance. Another major blood antigenic system MN was defined by Landsteiner. Antibodies raised in rabbits against the red cells of one individual agglutinated the red cells of some but not all other individuals. Antibodies raised against red cells obtained from the 'negative' individuals in this test caused erythrocyte agglutination of the individuals who were negative using the original antiserum. Hence two allelic antigens M and N and a subgroup MN, individuals having red cells carrying surface antigens recognized by both sets of antisera, were defined. It now turns out that the differences in MN antigenicity resides, at least in part, in the unique aminoacid sequences between residues 1 and 5 in glycophorin A. M individuals have the sequence

$$\text{ser}\,.\,\overset{*}{\text{ser}}\,.\,\overset{*}{\text{thr}}\,.\,\overset{*}{\text{thr}}\,.\ \text{gly},$$

while N individuals have the sequence

$$\text{leu}\,.\,\overset{*}{\text{ser}}\,.\,\overset{*}{\text{thr}}\,.\,\overset{*}{\text{thr}}\,.\,\text{glu}.$$

However, the *O*-glycans substituted onto the starred residues play an important role in the antigenicity, possibly by maintaining the two peptide sequences in a fixed conformation recognized by the specific antisera. Of particular significance in the glycophorin A structure is a stretch of aminoacids with neutral or apolar side chains: residues 73–95. This domain almost certainly inserts into the lipophilic bilayer of the membrane so that one end of the polypeptide extends from one side of the bilayer and the other domain is located on the opposite side. The amino-terminal, externally located domain is very rich in two basic types of carbohydrate chains. There are fifteen *O*-glycans consisting of a di, tri or tetrasaccharide linked through *N*-acetylgalactosamine residues to serine or threonine. The structures are similar to the *O*-glycans already discussed in containing the galactosyl $\beta 1 \rightarrow 3$ *N*-acetylgalactosamine sequence which may be sialylated further on one or both monosaccharides. These oligosaccharides are not unique to glycophorin but exist

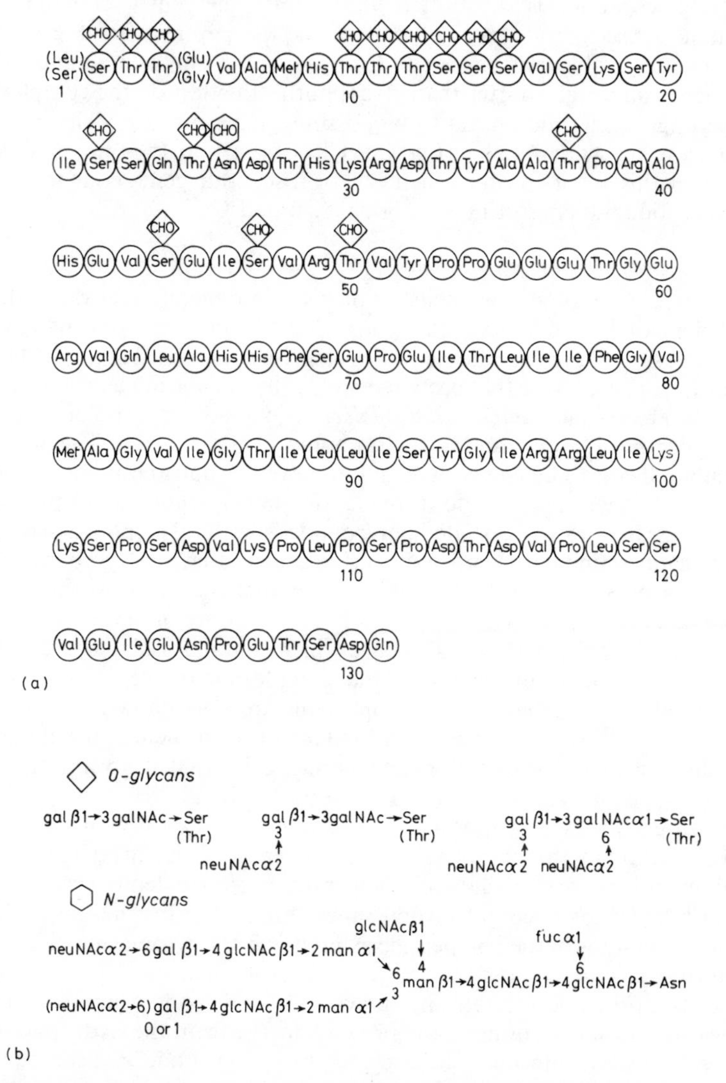

Fig. 2.19 (a) Primary structure of glycophorin A showing the glycosylated amino acid residues. (b) Generalized structures of the carbohydrate units.

in soluble glycoproteins such as fetuin, the principal glycoprotein of fetal blood. A single asparagine residue of glycophorin is glycosylated with more complex chains. These chains bear features in common with the *N*-glycans of soluble glycoproteins: in particular a core region of a β-mannosyl unit linked to chitobiose [20]. The β-mannose residue is substituted with two α-mannose units which form attachment points for the *N*-acetyllactosamine sequence. Terminal sialic acid and fucose residues are substituted on to these sequences. In addition *N*-acetylglucosamine is substituted $\beta 1 \rightarrow 4$ onto the central β-mannose unit, as in the hybrid chains of ovalbumin glycopeptides (see Fig. 2.7).

The minor glycophorins B and C appear to be similar in aminoacid sequence although shorter than glycophorin. In addition the complex-type oligosaccharide apears to be missing suggesting that deletion of the sequence with Asn ^{26} (Fig. 2.19) has occurred. It is not known whether this occurs post-synthetically or if separate genes code for the polypeptide moieties of glycophorins A, B and C.

2.6.2 The erythrocyte membrane

The integration of glycoproteins or proteins into membranes raises the problem of how a polypeptide consisting of many aminoacids with charged side chains can interact with a structure composed of lipids. The solution appears to be that proteins employ the same stratagem as lipids. Lipids are in fact *amphipathic molecules* containing a polar, often charged, head group and a hydrophobic fatty acid tail. The lipids accumulate into a bilayer with polar groups on the outside and fatty acid chains directed inwards to provide the impermeable hydrophobic interior. Membrane-associated polypeptides such as glycophorin have a stretch of aminoacids with apolar side chains that neatly spans the membrane. Several lines of evidence show that the carbohydrate-rich domain of glycophorin A extends on the outer surface of the erythrocyte while the carboxyl-terminal sequence which contains a high content of acidic aminoacids extends into the cytoplasmic space (Fig. 2.20). Although several membrane glycoproteins are now known to have a similar organization, studies of a second major constituent of the human erythrocyte membrane cautions against supposing that all glycoproteins conform to this simple model. The so-called 'Band 3' protein of the human erythrocyte membrane is the anion-transport system in these cells. Although the complete amino sequence of the protein is still unknown it is certain that the polypeptide moiety extends across the lipid bilayer not once but several times (Fig. 2.20). Furthermore, the amino-terminus of the polypeptide extends into the cytoplasm while the carboxyl-terminal region bearing glycan is exposed at the outer surface. The same principle as established previously for glycophorin is followed however in that portions of the protein emerge from the hydrophobic interior of the membrane. These segments of polypeptide are believed to interact with other proteins. There is good evidence, for example, that a globular protein called *anchyrin* binds selectively to the amino-terminal

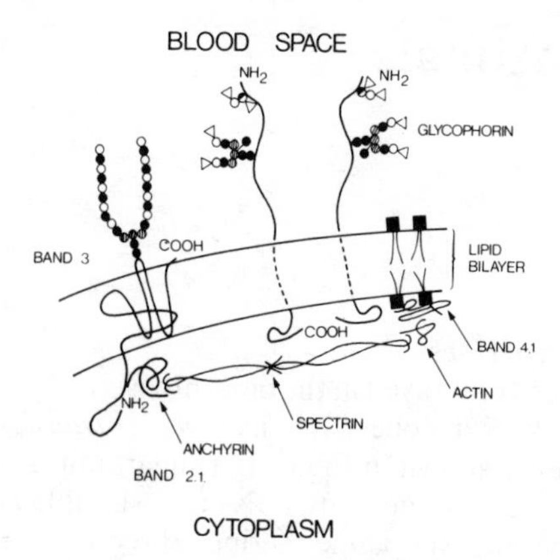

Fig. 2.20 Model of the arrangement of some proteins in the human red cell membrane. Monosaccharides are depicted by: ●, glcNAc; ⊘, man; ○, gal; ◐, galNAc; ▽, neuNAc. Only one of fifteen *O*-glycans are shown on glycophorin. Phospholipid head groups are shown (■). (Adapted from an original drawing of Dr Minoru Fukuda [21].)

sequence of the Band 3 protein. In turn anchyrin appears to interact with a fibrillar protein of the erythrocyte cytoplasm, *spectrin*. A meshwork of spectrin underlies the erythrocyte membrane and may contribute rigidity to the structure, maintaining the characteristic biconcave configuration of the erythrocyte. Cell shape may also be maintained by the complement of actin present on the cytoplasmic face of the membrane. Spectrin appears to bind to actin or actin tetramers at a site removed from that engaging anchyrin. Actin itself is possibly anchored to the membrane through interaction with another protein component of erythrocytes, the so-called 'Band 4.1' protein. Band 4.1 protein has been shown to have some affinity for the polar head groups of phospholipids and hence may stablize interactions between actin tetramers and the inner side of the erythrocyte membrane [21].

These findings offer intriguing insights into the possible ways by which a simple membrane may be 'toughened' by a muscular apparatus in order for a cell to adopt an unusual and energy demanding conformation and have implications for the maintenance of cell morphology and mobility in general [22]. Some of these aspects are considered briefly in a later chapter.

3 Biosynthesis

3.1 Glycosyl transferases

The enzymes directly involved in the biosynthesis of *N*- and *O*-glycans of glycoprotein are known collectively as *glycosyl transferases*. The basic reaction catalysed is shown in Fig. 3.1: a sugar unit is transferred from an activated, high energy derivative (X) to a suitable acceptor, Y. The acceptor may be an asparagine, serine, threonine or hydroxylysine residue in a polypeptide or a carbohydrate sequence in an incomplete glycan. The reaction may be repeated, e.g. in the synthesis of glycogen where glucose units are transferred from a high energy intermediate to the non-reducing end of the growing polysaccharide chains (Y). Glycolipids are formed similarly by transfer to the primary hydroxyl group of ceramide followed by sequential additions of further monosaccharides.

In general, individual glycosyl transferases exhibit very exact specificities towards the nature of groups X and Y. In fact these enzymes are among the most discriminating biologically active proteins known which has often retarded progress in their detection and characterization. Nevertheless we now know the action patterns of several glycosyl transferases in some detail and many are available in homogenous form.

3.2 The donors

The immediate donors of monosaccharides in glycosylation reactions are the α- or β-glycosyl esters of nucleotides, called simply *sugar nucleotides* (Fig. 3.2). The nucleotide moiety may be uridine, guanosine or cytidine which exist in sugar nucleotides as pyrophosphate esters except in sialic acid derivatives in which a phosphodiester linkage between the ribose moiety of the nucleotide and the sialic acid is found. These sugar nucleotides are the major ones involved in protein glycosylation. However, other derivatives occur in nature: for example glucose in plants is carried by UDP, adenosine diphosphate (ADP) or GDP and in bacteria by UDP, thymidine diphosphate (TDP) and CDP and all are involved in the biosynthesis of complex carbohydrates.

The sugar nucleotides of glucose, mannose and *N*-acetylglucosamine are synthesized by specific synthetases using α-glycosyl phosphates and nucleotide triphosphates with elimination of pyrophosphate (Fig. 3.3a).

$$\text{X—sugar} + \text{Y} \rightarrow \text{sugar—Y} + \text{X}$$

Fig. 3.1 Reaction catalysed by glycosyl transferases.

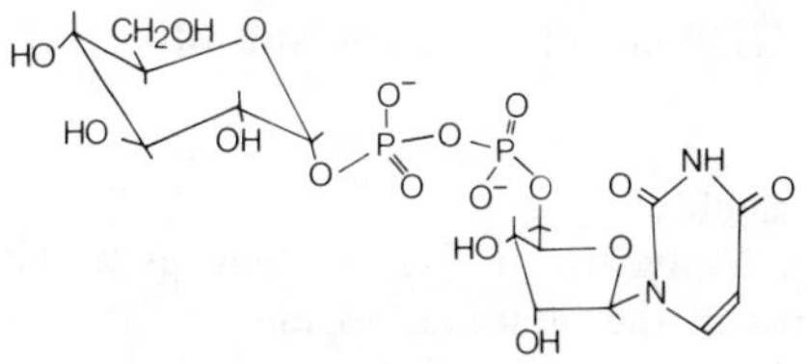

Uridine diphosphate glucose (UDPG)

Fig. 3.2 Sugar nucleotides.

Sugar	*Activated form*	*Anomeric form*
Galactose	UDP gal	α
Mannose	GDP man	α
N-acetylglucosamine	UDP glcNAc	α
N-acetylgalactosamine	UDP galNAc	α
Fucose	GDP fuc	β
N-acetylneuraminic acid	CMP neuNAc	β
Xylose	UDP xyl	α
Glucuronic acid	UDP glcUA	α

Although these reactions are freely reversible in the test tube, in practice they are rendered irreversible in cells by the rapid hydrolysis of pyrophosphate by pyrophosphatases. The sugar nucleotides formed undergo secondary enzymic reactions: UDPglc for example undergoes epimerization at C(4) to form UDPgal, carboxylation at C(6) to form UDPglcUA which may be decarboxylated to form UDPxyl. Similarly, UDPgalNAc is formed by C(4) epimerization of UDPglcNAc while GDPman undergoes reduction at C(6) to produce GDPfuc. In the latter case an inversion of configuration takes place: GDP-D-mannose is converted to GDP-L-fucose, and L-fucose is always found in glycoproteins.

Synthesis of activated sialic acids is different to the general reactions just described: (a) synthesis involves sialic acid (*N*-acetyl, *N*-glycolyl or *N*,*O*-polyacyl derivatives) and CTP to form the sugar monophosphate derivatives, with elimination of pyrophosphate; (b) unlike other sugar nucleotide synthetases, which are present in the cytoplasm, CMP sialic acid synthetase is found predominantly in the cell nucleus.

The monosaccharide moieties of sugar nucleotides are transferred to a suitable acceptor to form a particular glycosidic linkage. The synthesis of lactose is a simple illustration of these transfer reactions (Fig. 3.3b). *Lactose synthetase* or UDP-galactose : glucose galactosyl transferase catalyses step (b) and like all known glycosyl transferases is very specific

(a) glc-1-P + UTP ⟶ UDP-glc + PP

(b) UDPgal + glc ⟶ galβ1→4glc + UDP

Fig. 3.3 Biosynthesis of a sugar nucleotide and the disaccharide lactose.

for the donor; GDPgal, etc. cannot substitute for UDPgal in the reaction.

3.3 *N*-Glycan assembly

The structures of *N*-glycans of glycoproteins as we have seen are very diverse. Structures of the oligomannosidic, hybrid and complex types have been described in Chapter 2. In spite of this structural diversity biosynthesis of the *N*-glycans of most (perhaps all) glycoproteins, involves a common lipid intermediate based on the polyisoprenoid, *dolichol* [23, 24]. This pathway has been conserved remarkably during evolution [24].

3.3.1 Lipid intermediates

The role of polyisoprenoids in glycosylation was first demonstrated in the biosynthesis of certain bacterial cell wall polymers: the polysaccharides carrying type-specific antigenically active carbohydrate sequences studied by Phillips Robbins and his colleagues and the structural peptidoglycan of the cell wall studied by Jack Strominger and his group. Later the eminent Argentinian biochemist Luis Leloir produced convincing evidence that similar isoprenoids are involved in the glycosylation of proteins.

The general structure of the active polyisoprenoids is shown in Fig. 3.4. The value of *n* varies, but in mammalian tissues is 17 to 22. The

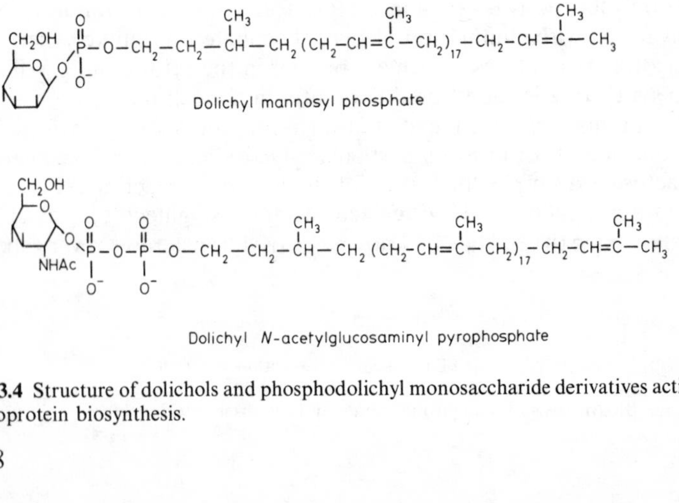

$$H{+}(CH_2-\overset{CH_3}{\overset{|}{C}}=CH-CH_2)_n-[CH_2-\overset{CH_3}{\overset{|}{C}H}-CH_2-CH_2]-OH$$

Isoprene unit   Saturated isoprene unit

Dolichol

$$CH_2OH \ldots O-\overset{O}{\overset{||}{P}}(O^-)-O-CH_2-CH_2-\overset{CH_3}{\overset{|}{C}H}-CH_2(CH_2-CH=\overset{CH_3}{\overset{|}{C}}-CH_2)_{17}-CH_2-CH=\overset{CH_3}{\overset{|}{C}}-CH_3$$

Dolichyl mannosyl phosphate

$$CH_2OH \ldots NHAc \ldots O-\overset{O}{\overset{||}{P}}(O^-)-O-\overset{O}{\overset{||}{P}}(O^-)-O-CH_2-CH_2-\overset{CH_3}{\overset{|}{C}H}-CH_2(CH_2-CH=\overset{CH_3}{\overset{|}{C}}-CH_2)_{17}-CH_2-CH=\overset{CH_3}{\overset{|}{C}}-CH_3$$

Dolichyl *N*-acetylglucosaminyl pyrophosphate

Fig. 3.4 Structure of dolichols and phosphodolichyl monosaccharide derivatives active in glycoprotein biosynthesis.

terminal isoprene unit is saturated and carries a primary hydroxyl which in the active form is phosphorylated. This reaction can be done by a specific CTP-dependent kinase. When extracts of a large variety of mammalian cells, as well as insects and plants and some lower eukaryotes (such as yeasts, other fungi and protozoa), are incubated with dolichol phosphate and GDP-mannose or UDP-glucose, UDP or GDP is released and sugar derivatives soluble in organic solvents are formed. The sugar residues are linked to the polyisoprenoid by a phosphodiester bridge in the β-configuration (Fig. 3.4). Since the α- forms are present in the sugar nucleotides, formation of these dolichyl sugar phosphates involves an inversion on C(1). By contrast *N*-acetylglucosamine-1-phosphate is transferred from UDP-*N*-acetylglucosamine to dolichol phosphate with elimination of UMP (Fig. 3.4). Hence the α-configuration of the sugar unit is preserved in dolichyl pyrophosphate *N*-acetylglucosamine.

The simple derivatives shown in Fig. 3.4 are not the only lipophilic compounds synthesized by cell extracts. Leloir made the fundamental discovery of dolichol phosphate derivatives with unusual solubility properties and showed the presence of oligosaccharides attached by a pyrophosphate bridge to the primary hydroxyl of the dolichol. This finding has been confirmed in many cells and tissue extracts and in time the complete sequence of reactions involving dolichol phosphate and sugar nucleotides has been resolved (Fig. 3.5). In addition, certain aspects of the regulation of this unique metabolic pathway are now established.

The levels of dolichol phosphate in cells and tissues is controlled by at least two enzymes: the kinase previously mentioned and a phosphatase. Reaction of UDPglcNAc with dolicholphosphate is catalysed by a specific enzyme producing the pyrophosphate derivative containing one *N*-acetylglucosamine linked to dolichol. This residue then forms the attachment point for enzymic transfer from UDPglcNAc of a second *N*-acetylglucosamine in β1–4 linkage, followed by attachment of a β-mannosyl residue to C(4) of this new unit. The donor in this reaction is GDP-mannose. The next steps are the transfer of eight mannose residues in α-linkage and three glucose units to form a unique branched structure [23, 24]. At present these mannosyl transferases have not been purified. However it seems clear from kinetic experiments using whole cells that the attachment of the α-mannosyl units is a highly ordered process (Fig. 3.6) and perhaps conserved during evolution. The mannosyl α1 → 3 is linked first to the central β-mannose unit followed by the mannosyl α1 → 6 unit. The branch containing the former α-mannose residue is then elongated by two mannosyl α1 → 2 units before assembly of the polymannose chain on the second branch. Again the α1 → 3 linked arm is elaborated before the α1 → 6 linked arm and glucose is finally added [25]. It is probable that these steps require separate α-mannosyl transferases and three glucosyl transferases.

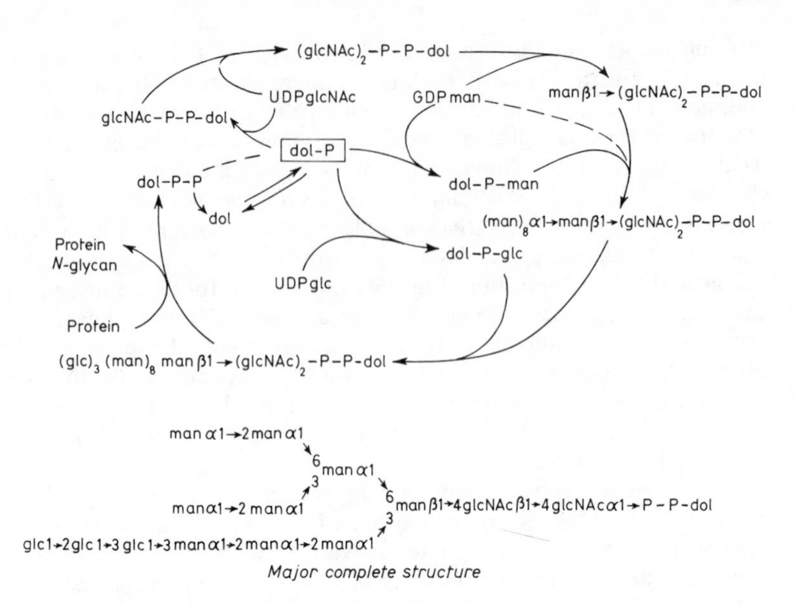

man α1
6
man β1→4 glcNAc β1→4 glcNAc α1→P–P–dol
3
glc1→2glc1→3glc1→3man α1→2man α1→2man α1

Minor variant structure

Fig. 3.5 The lipid cycle and the structure of the terminal product.

Cells that are blocked specifically in dolichyl mannose phosphate synthesis produce only a smaller variant lipid intermediate containing four α-mannosyl residues (Fig. 3.5) that is a very minor component of normal cells. The inference then is that these units are transferred from GDP-mannose while the other four α-mannose residues derive from dolichol mannose phosphate [26, 27]. The other steps in assembly are the attachment, probably from dolichol glucose phosphate, of three glucose units to one branch of the mannose-rich structure.

The final major product (Fig. 3.5) is the direct precurser of *N*-glycans of glycoproteins; the oligosaccharide moiety of the lipid intermediate is transferred *en bloc* to asparagine residues in a polypeptide acceptor. Dolichol pyrophosphate is eliminated and probably returns into the dolichol pool by dephosphorylation (Fig. 3.7).

The enzyme responsible for this transfer has been detected in many tissues active in glycoprotein biosynthesis. The enzyme is remarkably universal: enzyme preparations from one source show activity towards a variety of polypeptide acceptors prepared from the same tissue or from a completely different source. Even more remarkably, small peptides

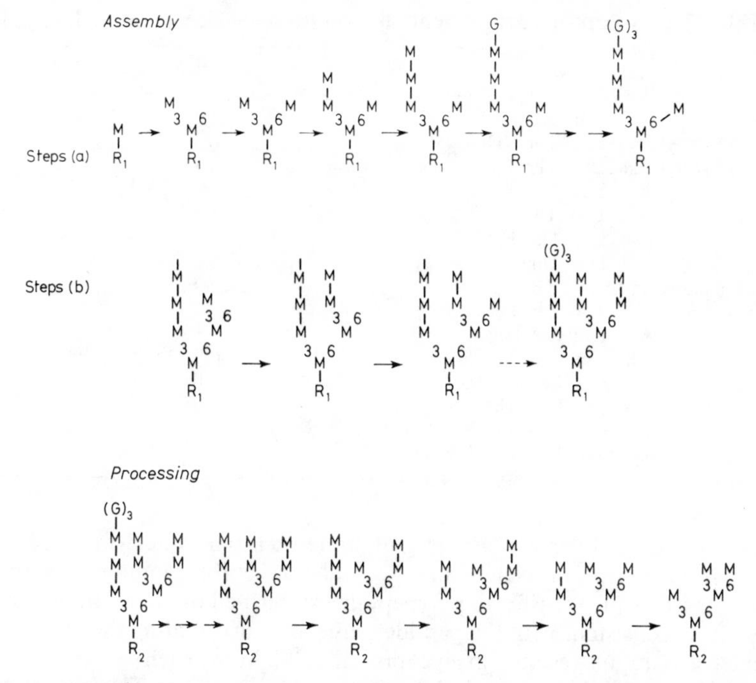

Fig. 3.6 Assembly of oligomannosidic lipids and patterns of processing in Chinese hamster ovary CHO cells. R_1 = glcNAc $\beta 1 \rightarrow 4$ glcNAc pyrophosphate dolichol; R_2 = glcNAc $\beta 1 \rightarrow$ glcNAc . Asn. Mannose (M) and glucose (G) residues are in the sequence shown in Fig. 3.5. (a) Minor pathway; (b) Major pathway.

containing asparagine are able to act as acceptors in the reaction [28, 29]. For example when the acceptor activity of *unfolded* (carboxymethylated) ribonuclease or α-lactalbumin is compared with the synthetic peptides based on the sequences around the asparagine residues glycosylated in the intact proteins, the peptides compare very well indeed (Table 3.1). However, not *all* peptides are active; only those having the sequence Asn . x . Ser or Thr, where x can be almost any aminoacid residue except proline and possibly aspartic acid, are active (Table 3.1). In order to function as acceptors, the asparagine and hydroxylaminoacid residues must have blocked amino- and carboxyl-terminal groups, and the asparagine residue is essential for acceptor activity, not surprisingly since this residue is glycosylated. Thus, the

$$(glc)_3\ (man)_9\ (glcNAc)_2\text{–P–P–dol} + \text{Asn–polypeptide} \rightarrow$$
$$(glc)_3\ (man)_9\ (glcNAc)_2\text{–Asn–polypeptide} + \text{dol–PP}$$

Fig. 3.7 Final step in glycosylation of asparagine residues in proteins.

Table 3.1 Acceptor requirements for oligosaccharide transfer [28, 29]

Synthetic peptides	*Relative activity* (%)
RNAase (Asn^{34}. Leu^{35}. Thr^{36}.)	100
N^{α}-acetyl Asn . Leu . Thr $NHCH_3$	473
N^{α}-acetyl Asn . Leu . Thr . Lys .	293
Ser . Arg . Asn . Leu . Thr . Lys .	120
N^{α}-acetyl Asn . Leu . Thr .	5
Asn . Leu . Thr . Lys .	1
Asn . Leu . Thr .	1
α-Lactalbumin (Asn^{45} . Gln^{46} . Ser^{47} .)	100
N^{α}-acetyl Asn . Leu . Ser . Leu .	18
Asn . Leu . Ser . Leu .	<1
Tyr . Asn . Leu . Thr . Ser . Val .	100
Tyr . Asp . Leu . Thr . Ser . Val .	0
Tyr . Gln . Leu . Thr . Ser . Val .	0
Tyr . Asn . Leu . Val . Ser . Val .	0
Tyr . Asn . Pro . Thr . Ser . Val .	0

analogous peptides containing glutamine or aspartic acid in place of asparagine are completely inactive. The acceptor requirements for oligosaccharide transfer to polypeptide established by these studies are entirely consistent with the peptide sequences surrounding the glycosylated asparagine residues in glycoproteins [30]. Many such sequences are now known and in all cases the sequence Asn . x . Ser or Thr is located internally in the glycosylated polypeptide chain. The inhibitory effect of proline is also predicted from a consideration of glycosylation sites in glycoproteins. The reason for this inhibition is not known but one suggestion is that hydrogen bonding can form between the amide of asparagine and the hydroxyl group of serine or threonine, and a proline residue prevents it. It may be that bonding is important in inducing a conformation suitable for substitution of the amide group by an incoming oligosaccharide; for example by modifying the ionization of the amide group.

Although oligosaccharide transfer to asparagine clearly has the tripeptide sequence as a necessary requirement, this is in itself not sufficient. Many proteins contain the required sequence that is, however, not glycosylated or only inefficiently glycosylated in the course of normal cellular biosynthesis. The pancreatic enzyme ribonuclease is an excellent example. Only about 20% of the protein molecules are glycosylated on asparagine 34 (see Table 3.1), whereas the remainder remain unglycosylated and serve as excellent acceptors *provided* the polypeptide moiety is first unfolded by appropriate chemical treatment (e.g. hydrolysis of disulphide bonds and carboxymethylation under denaturing conditions). Similarly α-lactalbumin contains two recognition tripeptides, Asn . Gln . Ser and Asn . Thr . Ser but the majority of these sequences are unglycosylated and only a small proportion are glycosylated at the first site. The denatured protein is readily glycosylated on either tripeptide sequence and the synthetic peptides based on these sequences are themselves excellent acceptors. The conclusions would then follow that (1) the tripeptide sequence must be sufficiently exposed on the outer surface of the protein molecule in order to be glycosylated during assembly in the cell; or (2) control mechanisms exist so that only those asparagines are glycosylated that occur in sequences destined to end up on the protein surface after biosynthesis and folding of the fully glycosylated polypeptide. It is quite clear that the bulky carbohydrate groups *always* exist on the outside of globular proteins and usually in peptide segments known as beta-turns. These conformations are presumably especially favourable for the inclusion of glycosylated sequences.

As mentioned in some circumstances the full assembly of the glucose-containing lipid intermediate can be prevented in whole cells. The formation of *N*-glycosylated proteins still occurs in such cells, however, and the question of the specificity of the glycosyl transferase catalysing oligosaccharide transfer to polypeptide then arises. It seems that several dolichyl pyrophospho-oligosaccharides can function as intermediates in this reaction. Even dolichyl pyrophospho-*N*-acetylglucosamine β1–4 *N*-acetylglucosamine is utilized by the enzyme for transfer of the chitobiose unit to synthetic peptides, although the efficiency of transfer is somewhat reduced [28]. This, together with the presence of the larger glucose-containing lipid intermediate as the major oligosaccharide linked to dolichol in cells under normal conditions, strongly suggests that the predominant pathway of *N*-glycosylation in cells is through transfer of this glucose-containing oligosaccharide to polypeptide. Direct evidence for this proposal has come from pulse-chase experiments with cells or virally-infected cells. In such studies it is clear that the oligosaccharide initially transferred to polypeptides of the cells or into the viral glycoproteins is identical in size and structure to that of the larger glucose-containing dolichol intermediate. In some as yet unexplained way, the glucose moiety of the fully assembled intermediate is an important factor controlling the efficient transfer to proteins, at least in mammalian cells [31].

3.3.2 Processing

Inspection of the oligosaccharide structure of the major lipid intermediate (Fig. 3.5) shows many features in common with known *N*-glycans of glycoproteins. The core region consisting of a β-mannosyl residue linked to a chitobiose unit is commonly found in *N*-glycans, as discussed previously. Furthermore the arrangement of α-mannosyl residues attached to this core region is identical in the lipid intermediate and, for example, the largest oligomannosidic unit of IgM (Fig. 2.8). There are a number of anomalies to be explained however. Firstly, the lipid-linked oligosaccharide contains three glucose residues while glucose is not commonly found as a constituent of asparaginyl-linked oligosaccharides. Secondly, we have to explain how the many other oligomannosidic structures of glycoproteins such as those present in ovalbumin (Fig. 2.6) arise. Finally, how are the complex *N*-glycans, for example of orosomucoid, assembled? It now seems probable that all of these structures arise from the product of transfer of the common glucose-containing dolichol derivative. The complete oligosaccharide of the lipid intermediate undergoes extensive modifications after transfer to polypeptide. 'These modifications occur in a complex series of reactions known as *processing*. Very soon after attachment to the polypeptide, the glucose residues are removed by highly specific α-glucosidase(s) [32]. α-Mannose residues are then removed by several specific α-mannosidases [33], apparently in a particular sequence [34] which, at least in some cases, is different to that of their assembly into the lipid intermediate (Fig. 3.6). In the course of processing (outlined in Fig. 3.6) many of the oligomannosidic glycans found as stable components of mature glycoproteins are formed, e.g. in immunoglobulin M at site (b) (Fig. 2.8). It is tempting to believe therefore that a controlled processing accounts for the array of oligomannosidic structures seen in glycoproteins, although a more complicated biosynthetic pathway could operate. For example, complete processing down to a five mannose-containing structure followed by resynthesis of the glycans with higher mannose contents. It is also possible that other pathways of processing occur to account for the different oligomannosidic structures found at site (a) in immunoglobulin in M for example (Fig. 2.8). At present nothing is known about the factors controlling processing and how the exact sequence of removal of mannose residues takes place according to a particular site of glycosylation on the same polypeptide. Such variation according to cell type or species could lead to discrete glycan chains that are characteristic of those cells, and contribute to the structural diversity of *N*-glycans.

3.3.3 Termination reactions

So far we have accounted for the assembly of *N*-glycans of the oligomannosidic type containing up to nine mannosyl units. How then is the second major class of *N*-glycans containing peripheral units of sialic acids, galactose and additional *N*-acetylglucosamine formed? Another problem is the control of attachment of fucose residues. It is an interesting point that the fucose attached to the chitobiosyl sequence in the core is not found on oligomannosidic glycans. Also, fucose is often substituted on to the gal $\beta 1 \rightarrow 4$ glcNAc structure in the non-reducing termini of complex glycans, as is sialic acid, but structures are not found in which a sequence is substituted by both sialic acid and fucose.

The starting point for assembly of these complex *N*-glycans is the structure containing five mannose residues (Fig. 3.8). Harry Schachter and his colleagues have identified a specific β-*N*-acetylglucosaminyl transferase that utilizes this substrate to transfer the monosaccharide unit of UDP-*N*-acetylglucosamine to the α-mannose unit joined in α-$1 \rightarrow 3$ glycosidic linkage to the core region sequence [35]. The enzyme is

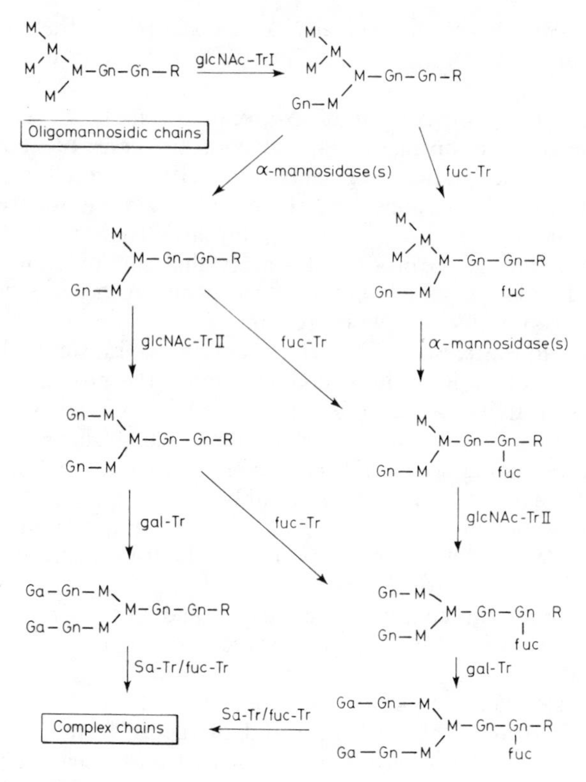

Fig. 3.8 Biosynthesis of *N*-glycans containing galactose. Tr, Transferase. (See Table 1.1 for explanation of abbreviations.)

known as *N-acetylglucosaminly transferase I* or *glcNAc-TrI*. Experiments with cell free extracts or the purified enzyme have shown that although other oligomannosidic glycopeptides or glycoproteins function as acceptors in this reaction, the preferred substrate is the structure with five mannose residues. Attachment of the *N*-acetylglucosamine unit triggers further processing of the oligosaccharide formed, by α-mannosidase action. This α-mannosidase(s) has been purified and hydrolyses the $\alpha 1 \rightarrow 3$ and $\alpha 1 \rightarrow 6$ glycosidic linkages. Mannosyl $\alpha 1 \rightarrow 2$ linkages are not cleaved and the enzyme is clearly separate from the enzymes involved in earlier processing reactions (see Fig. 3.6). The product is an asymmetric sequence containing two α-mannose units linked to the core region which is a substrate for another specific *N-acetylglucosaminyl transferase II* or *glcNAc-TrII*. Following reactions include the transfer of galactose units from UDP-galactose to C(4) of the *N*-acetylglucosamine residues to produce the *N*-acetyllactosamine sequence typical of many complex N-glycans. Recently [36] a galactosyl transferase capable of synthesizing the gal $\beta 1 \rightarrow 3$ glcNAc sequence at these points has been described as expected, since the sequence has been found in some *N*-glycans. The glcNAc-transferase I is therefore a key enzyme in conversion of oligomannosidic *N*-glycans into complex chains.

Another form of control of *N*-glycan assembly is seen in the biosynthesis of rhodopsin, the major glycoprotein constituent of the disc membrane of the rod outer segments in the retina. The cells responsible for biosynthesis of rhodopsin may contain low α-mannosidase activity and the structures of the *N*-glycan in rhodopsin [37] reflects this deficiency (Fig. 3.9). Although the product of *N*-acetylglucosaminyl transferase I is clearly formed in these cells and some processing takes place, no second *N*-acetylglucosamine residue is present.

The reaction catalysed by *N*-acetylglucosaminyl transferase I appears to trigger another important modification of the glycan. A fucosyl transferase activity has been detected in cell fractions that attaches a fucose residue in α1 → 6 linkage to the *N*-acetylglucosamine unit attached to asparagine [11, 35]. This transferase cannot act on the core structure $man_3glcNAc_2$.Asn but only when at least one *N*-acetylglucosamine residue is attached to an α-mannose unit. In other words, the *N*-acetylglucosaminyl transferase I must act before a suitable sequence is available for the specific fucosyl transferase (Fig. 3.8). This provides a good explanation for the structural data showing that such a linkage is not present in oligomannosidic *N*-glycans.

The final steps in complex *N*-glycan assembly are the attachment of sialic acids from CMP-sialic acid and of fucose from GDP-fucose. Several sialic acid derivatives are known to be present at the non-reducing termini of *N*-glycans in glycosidic linkage to the C(3) or C(6) of the penultimate galactose residues. Separate sialyl transferases appear to catalyse formation of these different glycosidic linkages. Further structural diversity then is generated by modification of the sialic acids for example by hydroxylation of the *N*-acetyl group to *N*-glycolyl or by the addition of *O*-acetyl groups that can occur after transfer to the glycan chain.

Comparison of the activities of a purified sialyl transferase catalysing

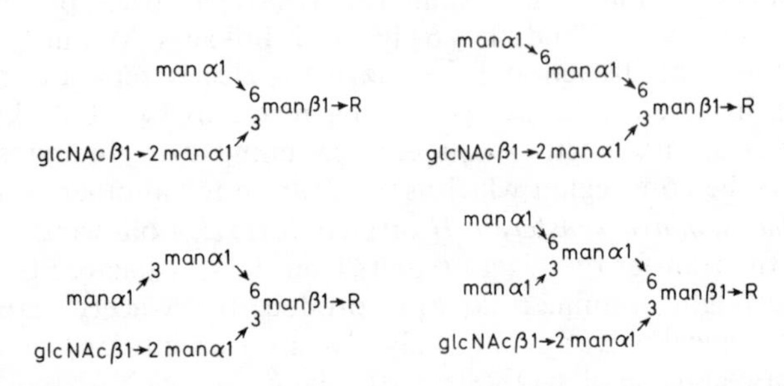

Fig. 3.9 The sugar chains of bovine rhodopsin. R = glcNAc β1 → 4 glcNAc . Asn.

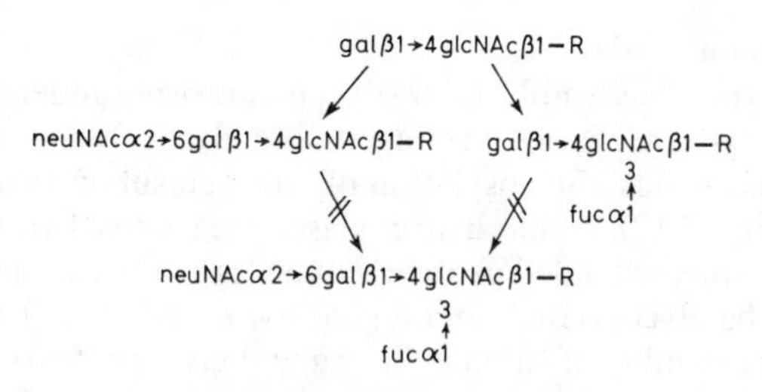

Fig. 3.10 Control of termination reactions in *N*-glycan biosynthesis. For R see Fig. 3.9.

formation of α2 → 6 glycosidic linkages to galactose has revealed [38] an interesting competitive interaction with a fucosyl transferase (Fig. 3.10). Terminal galactose β1 → 4 *N*-acetylglucosamine sequences are substrates for either sialic acid or fucose transfer but not both. The two reactions are mutually exclusive and provides an explanation for the recently determined [39] structure of the *N*-glycans of parotid α-amylase (Fig. 3.11). The fuc α1 → 3 glcNAc is also present on *N*-glycans of orosomucoid (Fig. 2.1) and other glycoproteins.

There are several major unsolved problems. To date no information has been obtained on the *N*-acetylglucosaminyl transferases attaching residues to the β-mannose unit in hybrid structures or to C(4) and C(6) of α-mannose residues in the highly branched complex *N*-glycans for example of orosomucoid (Fig. 2.11). It is highly likely however that these are separate activities and are different to transferases I and II. Secondly, we have no idea at present how the processing and termination reactions are controlled according to the location of asparagine residues in a polypeptide chain.

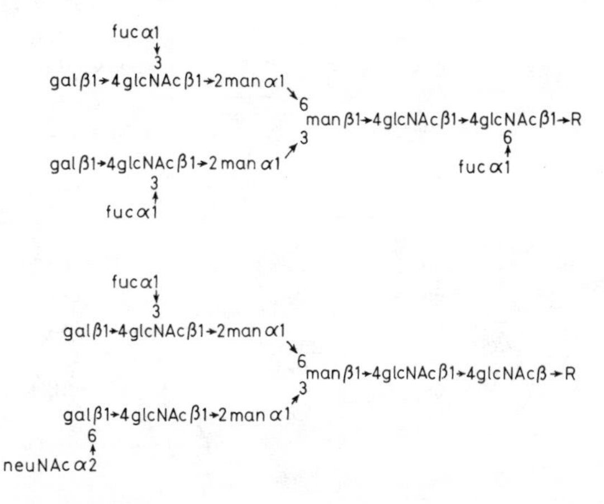

Fig. 3.11 Structures of *N*-glycans of parotid α-amylase [39].

3.4 *O*-Glycan assembly

The pathways of assembly of *O*-glycans are well understood for the short carbohydrate chains of collagens and glycoproteins secreted from submaxillary glands. Glycosylation of collagen subunits requires three enzymes (Fig. 3.12): lysine hydroxylase, a galactosyl transferase that recognizes rather extended peptide sequences around the hydroxylysine residues to be glycosylated, and a glucosyl transferase (Fig. 3.12) [40].

O-glycan assembly of submaxillary gland glycoproteins also involves direct transfer of sugars from nucleotide intermediates, catalysed by a multi-enzyme system. The first glycosyl transferase (Fig. 3.13, enzyme 1) establishes the linkage to serine or threonine residues using UDP *N*-acetylgalactosamine. The enzyme requires an extended peptide sequence surrounding the glycosylated aminoacid residues and most polypeptides are not substrates. An interesting exception is a major basic protein of nerve myelin. This protein is not normally glycosylated and the biological significance of its acceptor activity is not known. Synthetic peptides analogous to the sequence surrounding threonine 98 of myelin protein, the only glycosylated aminoacid in basic protein, have been used to define the peptide specificity of submaxillary gland galNAc-transferase (Table 3.2).

The essential features appear to be: threonine must be an internal residue in the sequence and a high proportion of proline residues facilitate glycosylation. Notice that the glycosylation is very specific for threonine 98. A threonine at site 95 is not an acceptor and the properties

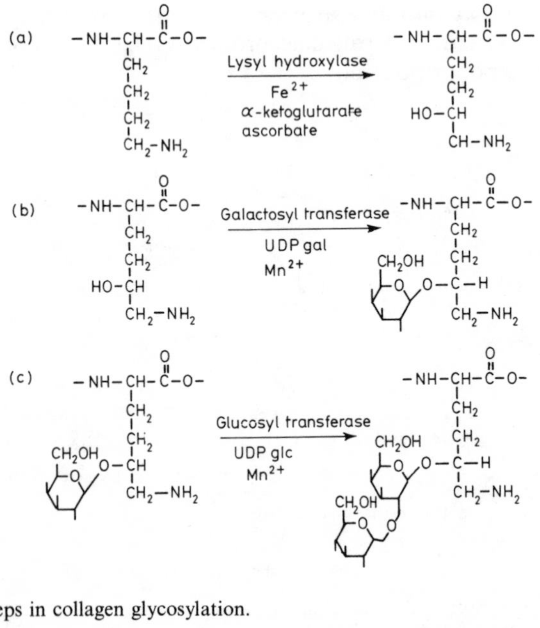

Fig. 3.12 Steps in collagen glycosylation.

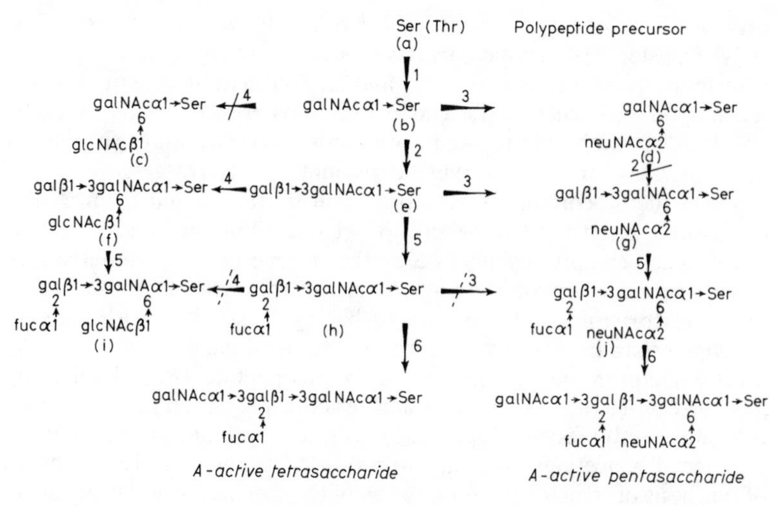

Fig. 3.13 Assembly of submaxillary gland glycoproteins. Disallowed reactions are indicated by full lines and partially blocked reactions by broken lines.

of the system seem to be similar to the rather exacting requirements of the *N*-acetylglucosaminyl transferase involved in glycosylation of asparagine residues. Presumably the high proline content of the polypeptides of the submaxillary gland glycoproteins ensures a proline-rich environment around each of the numerous glycosylated hydroxy-aminoacid residues in these molecules. The product of this reaction (Fig. 3.13b) is a substrate for two competing enzymes 2 and 3: a sialyl transferase adding sialic acid from CMPneuNAc to the *N*-acetylgalactosamine residue (Fig. 3.13d) and a galactosyl transferase (Fig. 3.13e). In tissues, e.g. sheep gland, with a high activity of the sialyl transferase the disaccharide accumulates as neuNAc α2 → 6 galNAc since the disaccharide is not a substrate for galactosyl transferase. The

Table 3.2 Requirements for glycosylation of a threonine residue. (From Young *et al.* [41])

Peptide (or protein)								*Relative acceptor activity*
Basic protein myelin								1.0
94	95	96	97	98	99	100	101	
Val.	Thr.	Pro.	Arg.	Thr.	Pro.	Pro.	Pro.	2.5
	Thr.	Pro.	Arg.	Thr.	Pro.	Pro.	Pro.	1.1
		Pro.	Arg.	Thr.	Pro.	Pro.	Pro.	1.2
			Arg.	Thr.	Pro.	Pro.	Pro.	1.1
				Thr.	Pro.	Pro.	Pro.	0.1
				Thr.	Pro.	Pro.		0
				Thr.	Pro.			0

product (Fig. 3.13e) is a substrate for three glycosyl transferases. The sialyl transferase (enzyme 3) produces a trisaccharide (g); a fucosyl transferase produces structure (h) and an *N*-acetylglucosaminyl transferase produces structure (f). The fucosyl transferase can act also on the sialylated trisaccharide (g) and both fucosylated structures (h) and (j) are substrates for an *N*-acetylgalactosaminyl transferase.

These fucose-containing structures will be recognized from earlier discussion as the blood group H determinant and the *N*-acetylgalactosaminyl transferase as the *A* gene enzyme conferring A blood group activity on the *O*-glycans. The third transferase active on structure (e) (Fig. 3.13) appears to be highly specific. Although *N*-acetylglucosamine is transferred from UDPglcNAc to the *N*-acetylgalactosamine residue linked to polypeptide (Fig. 3.13f), the galactose unit is essential for transferase action [42]. Even substitution of the galactose by fucose (Fig. 3.13h) drastically inhibits the subsequent transfer of *N*-acetylglucosamine to produce (i). The preferred pathway of synthesis of structure (i) is by action of the specific fucosyl transferase on substrate (f). Finally, the mutually competitive nature of fucosylation and sialylation reactions noted earlier is also displayed in another manner here. Transfer of fucose to a galactose residue in structure (h) greatly inhibits the subsequent transfer of sialic acid to an adjacent monosaccharide to produce structure (j) [43]. The preferred pathway for synthesis of this tetrasaccharide is therefore:

$$a \rightarrow b \rightarrow e \rightarrow g \rightarrow j.$$

These important experiments provide a convincing explanation for the finding of *only* the disaccharide neuNAc $\alpha 2 \rightarrow 6$ galNAc in a tissue like sheep glands, relatively poor in the specific galactosyl transferase activity in the presence of a very active sialyl transferase activity, and the production of longer glycans terminated by blood group active determinants in glycoproteins secreted by pig submaxillary gland. The relative activities of the sialyl and galactosyl transferases suggests a control mechanism by which the quantity of longer *O*-glycans can be regulated. Possibly formation of the *N*-acetylglucosamine $\beta 1 \rightarrow 6$ linkage triggers even more elaborate extensions of the glycan to form (e.g. the core region of the human blood group active products of human secretory cells (Fig. 2.16)) by attachment of other monosaccharides to the terminal galactose of the initial product (f) of the glcNAc transferase reaction.

The synthesis of the *O*-glycans present in glycophorin, fetuin (a major glycoprotein synthesized by fetal liver) and many other glycoproteins (see Fig. 2.19) may proceed along similar metabolic pathways to those just described. A specific sialyl transferase forming the sequence NeuNAc $\alpha 2 \rightarrow 3$ gal has been purified from pig liver and other tissues [43, 44] explaining the presence of the trisaccharide neuNAc $\alpha 2 \rightarrow 3$ gal $\beta 1 \rightarrow 3$ galNAc (see Fig. 2.19). However, control of activity of the enzyme forming the NeuNAc $\alpha 2 \rightarrow 6$ galNAc linkage to produce tetrasaccharide containing two sialic acids (see Fig. 2.19) is unclear at present. No activity analogous to enzyme 3 (Fig. 3.13) has yet been found in liver, the source of fetuin and evidently the sialyl transferase requires a different acceptor oligosaccharide [44]. The monosialosyl trisaccharide *O*-glycan (see Fig. 2.19) may be an essential intermediate in assembly of the tetrasaccharide *O*-glycan.

3.5 Cellular sites of glycosylation

So far we have considered the properties of specific glycosyl transferases in the test tube. Biosynthesis of glycoproteins is a highly ordered process within the cell; however, and we now need to consider the temporal flow of nascent glycoproteins during assembly and the cellular location of glycosylation reactions. Glycoproteins are constituents of most, perhaps all, membrane systems within the cell and most but not all secreted proteins are glycoproteins. However carbohydrate probably is not a necessary signal for secretion since proteins such as procollagen continue to be secreted in cells blocked in glycosylation, either genetically by a mutation in the pathways leading to glycosylation, or by various inhibitors of protein glycosylation. The most useful inhibitor of this kind is *tunicamycin* which prevents synthesis of dolichyl *N*-acetylglucosamine pyrophosphate, the first step in *N*-glycan assembly. However, recent studies discussed in Chapter 4 do suggest strongly that prevention of protein glycosylation often induces the appearance of glycoproteins in inappropriate places within the cell or leads to secretion of glycoproteins that, in normal circumstances, reside intracellularly. The idea is slowly forming that an ordered pattern of glycosylation leads to the directed flow of glycoproteins from the sites of polypeptide synthesis to the proper terminal sites. It is important therefore to consider where in the cell protein glycosylation occurs.

There seems little doubt that the polypeptide moieties of glycoproteins are produced by ribosomes bound to the membrane system of the endoplasmic reticulum, the 'rough' ER. By feeding cells active in glycoprotein synthesis with radioactive monosaccharide the attachment of *N*-acetylglucosamine and mannose to polypeptides in the rough ER can be readily shown by radioautography of cell sections or by subcellular fractionation. The pioneer in these methods is Claude Leblond of McGill University. Similar methods have shown that fucose, galactose, and *N*-acetylmannosamine, a direct and specific marker for sialic acids, are added in smooth membranes particularly the Golgi membranes. These observations can be confirmed by direct assays of specific glycosyl transferases in subfractions purified from disrupted cells. For example, the transferases responsible for assembly of the terminal sialyl. galactosyl. *N*-acetylglucosamine sequences of *N*-glycans, as well as fucosylation, are purified with Golgi membranes. Recently specific antibodies against galactosyl transferase were shown to stain, in whole cell mounts, a dense jucstanuclear cap which was interpreted to be part of the Golgi complex [45]. By contrast, dolichol-stimulated transferases are located in rough ER preparations. A reasonable scheme of glycoprotein biosynthesis taking these findings into account is incorporated in Fig. 3.14.

The nascent polypeptides carry a 'signal' peptide sequence according to the model popularized by Blobel and Dobberstein [46] which serves to fix the ribosomes to the endoplasmic reticulum membranes, an association stabilized by specific ER membrane proteins with affinity for some feature of the ribosome. Translation of mRNA continues with *vectorial discharge* of the nascent polypeptide into the cisternal space of the rough ER. At an early stage the signal peptide is cleaved by a specific proteolysis, folding begins and the exposed segments of the polypeptide becomes available for glycosylation. The glycosylation is confined to dolichol-mediated reactions, that is to say transfer of the oligosaccharide $(glc)_3(man)_9(glcNAc)_2$ to exposed asparagine residues in the correct peptide sequences. The oligosaccharide transferase is tightly bound to the membrane system [47]. Soluble glycoproteins are discharged entirely into the cisternal space upon completion of mRNA translation while glycoproteins destined to remain as membrane components remain integrated into the ER membrane by association through hydrophobic peptide segments such as that established for glycophorin.

The next stage in maturation is conversion of the 'rough' ER into ribosome-free 'smooth' membranes by disassociation of the ribosomes from the ER membranes, or the lateral migration of glycosylated proteins into areas of the ER deficient in the ribosome-binding proteins of the ER membrane. Transport of newly synthesized glycoproteins into the Golgi apparatus appears to be facilitated by the formation of vesicles coated on the cytoplasmic side by the protein clathrin [48]. These *coated vesicles* have been previously implicated in endocytosis of extracellular

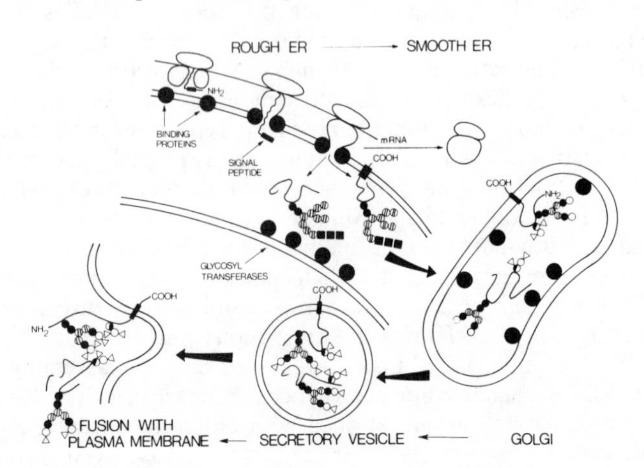

Fig. 3.14 A hypothetical scheme for glycoprotein biosynthesis. glcNAc (●), man (⊘), gal (◐), sialic acids (▽), galNAc (○), glc (■).

materials into the cell and a role in intracellular transport is now becoming clearer. The Golgi membranes in turn gives rise to various vesicular compartments including lysosomes and secretory vesicles. Further maturation of *N*-glycans occur at this stage since, as mentioned, the glycosyl transferases responsible for the further assembly and termination of *N*-glycans are located in the Golgi membranes. Furthermore, the processing glycosidases (glucosidases, mannosidases) are purified with smooth membrane fractions. Other events such as phosphorylation of *N*-glycans take place and appear to be important in directing the flow of glycoproteins specifically into lysosomes (see Section 4.3.2). The Golgi-derived vesicles, again coated with clathrin and equipped to mediate exocytosis of glycoproteins, then fuse with the plasma membrane when biosynthetic products are either secreted or remain as integral components of the surface membrane. Note that the topology of fusion confers on the plasma membrane an asymmetry that is observed by experiment. Thus, the glycosylated peptide segments that exist within the cisternale of the ER appear, after fusion, on the outer surface of the cell.

This simple discussion of very complex events cannot, of course, readily account for special problems such as the biosynthesis of Band 3 glycoprotein. As one possibility it may be that the signal peptide of some nascent polypeptides remains integrated into the rough ER membranes permanently. The elongating nascent polypeptide may wind across the ER membrane several times during elongation and ultimately becomes glycosylated on carboxyl-terminal segments.

The formation of *O*-glycans appears to be a relatively late event in biosynthesis [49]. The addition of monosaccharides adjacent to polypeptide, sialic acids and the conversion of *N*-acetylneuraminic acid by hydroxylation and acetylation reactions are reactions associated with smooth membranes or Golgi membranes. These different subcellular sites of core *N*-glycan assembly and terminating reactions in *N*-glycan biosynthesis and *O*-glycan assembly may be relevant to the involvement of basically different models of assembly. The close association of nascent polypeptides with the rough ER membranes may necessitate a role for lipid-linked intermediates while later events involving mature glycoproteins can take place by more conventional means.

3.6 Control of glycan biosynthesis

3.6.1 Metabolic control

A quiescent cell, such as a resting lymphocyte or a serum-starved fibroblast, maintains a basal level of protein synthesis and protein glycosylation and is stimulated in both capacities for example by application of a mitogen or serum. How does a cell such as a lymphocyte produce, upon stimulation, a massive increase in glycosylation potential accompanying the biosynthesis of immunoglobulins? Similarly, how do hormone-sensitive cells such as thyroid cells mobilize metabolic pathways to facilitate production of the specific glycoprotein end products, e.g. thyroglobulin? There are several possibilities. Perhaps the concentration of glycosylating enzymes and active sugar intermediates are saturating under resting conditions in cells. Glycosylation of the increased numbers of nascent polypeptide chains could then be accommodated. The available evidence, by contrast, suggests that glycosylation reactions are regulated by the cellular concentrations of dolichol and dolichol phosphate. Dolichol phosphate synthesis shares a common pathway with cholesterol and ubiquinone (Fig. 3.15) and the enzyme HMGCoA reductase catalysing formation of mevalonate (Fig. 3.15, step 2) appears to be an important control point for biosynthesis of both compounds. An inhibitor of this enzyme, 25-hydroxycholesterol, depresses cholesterol and dolicholphosphate synthesis and reduces protein glycosylation at least in some cells [50]. Similar inhibitions are produced by *compactin*, a potent inhibitor of HMGCoA reductase from *Penicillum citrinum* [51]. Interestingly, compactin fed to developing sea urchin embryos blocks a specific gastrulation stage and addition of exogenous dolichol returns development to normal, probably by a direct effect on glycosylation. In liver excess cholesterol depresses synthesis of HMGCoA reductase, but this may not be the primary control point in hepatic cholesterogenesis, as conversion of farnesyl pyrophosphate into cholesterol (Fig. 3.15, step 4) is also inhibited. For this reason there is a greater flow from farnesyl pyrophosphate into dolichol phosphate (Fig. 3.15, step 5) and ubiquinone (Fig. 3.15, step 8), and protein glycosylation is stimulated [52]. In liver under normal conditions the flow of precursors into dolichol phosphate is only about a hundredth of the flow into cholesterol. Therefore, sufficient farnesyl phosphate is produced even when HMGCoA reductase is partially inhibited, to saturate the pathways into dolichol phosphate.

Taken together these results point out an important metabolic control of protein glycosylation at the level of dolichol phosphate production. Since the dolichol phosphate pool is rate-limiting for glycosylation, additional controls probably also exist in the conversion of dolichol pyrophosphate (Fig. 3.15, step 9) to dolichol phosphate by specific

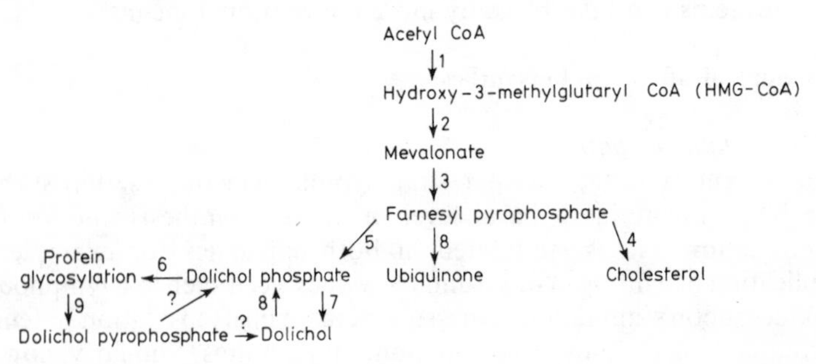

Fig. 3.15 Pathways of cholesterol and dolichol biosynthesis. CoA, coenzyme A.

phosphatases and the CTP-dependent kinase acting on a separate pool of free dolichol or dolichyl esters.

3.6.2 Genetic control

Genetic control of glycoprotein biosynthesis stems from two major areas: (1) translation from the genes by well-known template mechanisms and production of polypeptides having the appropriate aminoacid sequence (and perhaps conformation) about certain aminoacid residues capable of recognition for glycosylation; (2) the availability within the cell of a complex system of enzymes responsible for synthesis of activated forms of sugars, for transfer of those sugars to acceptors and for processing.

As we have seen, the assembly of glycans requires a multi-enzyme complex: in the build-up of the oligosaccharide sequences the product of one reaction becomes the substrate for a separate enzyme or several enzymes acting competitively. The diversity of carbohydrate structure in glycoproteins derives from this metabolic interplay. Of course, these enzymes are themselves direct products of genes and one obvious form of genetic control is in their synthesis. Direct evidence for such a genetic control comes from the properties of mutant cell lines which will be described later but the basic principles are perhaps best illustrated by consideration of the genetic control of biosynthesis (Fig. 3.16) of the major ABO(H) blood group antigens [53].

The A and B determinants as defined by antibodies recognizing tetrasaccharides of unique structure that may be attached to a variety of carriers are of course synthesized by specific glycosyl transferases, the presence of which in the cell is dictated by the genotype of the individual. The necessary preliminary step in assembly of A and B determinants is the attachment of a fucose residue to the penultimate galactose of the precursor. This enzyme is present in all individuals except for rare exceptions carrying the Bombay genotype discussed earlier. The antigen formed by addition of fucose is defined serologically as the H determinant and is the product of the enzyme specified by the *H* gene. The rare *h* gene of the Bombay individuals either does not code for an active enzyme or one of such weak activity that subsequent modifications of the oligosaccharide are effectively blocked during normal conditions of biosynthesis. The product formed by the activity of H transferase then is recognized by the enzymes specified by the *A* and *B* genes. The transferase controlled by the *A* gene is a UDP-*N*-acetylgalactosamine: *O*-α-L-fucosyl (1 → 2) D-galactose α-3-*N*-acetylgalactosaminyl transferase. Individuals of *A* genotype produce the enzyme and hence complex carbohydrates terminating in these sequences. By contrast *B* individuals produce the enzyme, UDP-galactose: *O*-α-L-fucosyl (1 → 2) D-galactose α-3-galactosyl transferase specified by the *B*-gene. Bombay individuals carry functional *A* and *B* genes, produce the active A or B enzymes but fail to produce A and B antigens because the necessary preliminary step is missing due to an inactive *H*

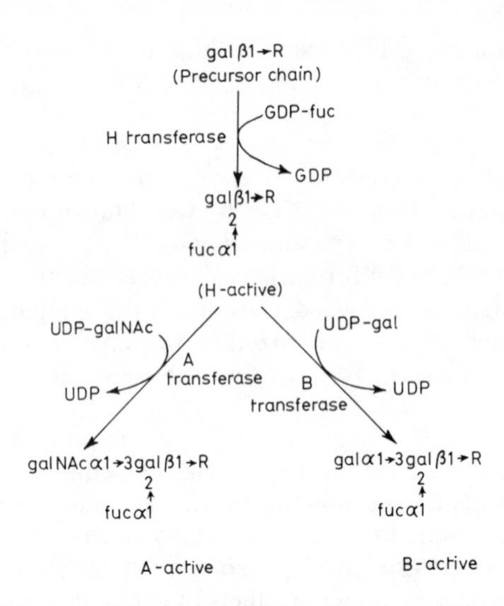

Fig. 3.16 Biochemical pathways for assembly of A, B and H structures. R = → 3 (or 4) glcNAc linked to oligosaccharides of different structures that do not contribute to ABO(H) blood group antigenicity as far as is known.

gene or *H*-gene product. Individuals carrying both *A* and *B* genes produce both A and B transferases and synthesize carbohydrate chains terminated in sequences recognized as A and B determinants (Fig. 3.16).

The *A* and *B* genes are thought to be allellic, an unusual situation since they code for enzymes with qualitatively different specificities, whereas usually the products of allelic genes show only quantitative differences in biochemical action. The third allele in the ABO system is the gene presumed to be silent in the sense that it does not give rise to a glycosyl transferase acting on the product of the *H*-gene. In blood group O individuals the H-active structure is not modified further into a serologically recognized form.

3.6.3 Modulators of glycosyl transferases

Another type of genetic control over glycosylation reactions is shown by a galactosyl transferase responsible for lactose synthesis (Fig. 3.3). In the mammary gland the activity resides in a complex of a catalytic subunit, a glycoprotein of molecular weight 50 000 and α-lactalbumin of molecular weight 14 000. These subunits form a 1:1 molecular complex and catalyse transfer of galactose from UDP-galactose to glucose (Fig. 3.3b). In tissues not making lactose however, the catalytic subunit exists alone and now transfers galactose from UDP-galactose to terminal *N*-acetylglucosamine residues of glycoproteins, or to free *N*-acetylglucosamine to form the gal β1 → 4 glcNAc sequence present in many *N*-glycans. The modulator subunit, α-lactalbumin promotes the binding of glucose to galactosyl transferase through a series of synergistic equilibria, so that the K_m is lowered from 2 M to 2 mM or less. Hence, in mammary tissue uniquely containing α-lactalbumin, lactose synthesis is catalysed effectively at physiological concentrations of glucose. Thus, α-lactalbumin is a non-catalytic 'modulator' of a galactosyl transferase.

It will be interesting to see if other examples of this type of genetic control of enzyme action exists, e.g. in the glycosyltransferases responsible for attachment of the same monosaccharide to closely similar substrates.

4 Function

4.1 Polypeptide processing

Many polypeptides or proteins are synthesized as giant precursors that are subsequently cleaved by specific proteolysis into mature, biologically active products. The glycans of glycosylated polypeptides are bulky substituents that would appear to play important roles in controlling fragmentation of these precursors in appropriate places to produce efficiently the active components. One example is described.

The pituitary produces a bewildering array of biologically active peptides, for example adrenocorticotropin, melanotropin and endorphins. A rational scheme for production of such an array has come from the identification of a single precursor, isolation of the mRNA from pituitary intermediate lobes and gene cloning. The cDNA sequence allows the segments coding for each active polypeptide to be arranged in sequence. The segments are flanked by pairs of basic aminoacids which may act as signals for highly specific cleavage by trypsin-like pituitary peptidases. In the *pars distalis* of the pituitary, fragmentation is stopped at the stage of the *N*-terminal fragment bearing the *N*-glycans, corticotropin and β-lipotropin. In the *pars intermedia* the last two fragments are cleaved extremely rapidly into β-endorphin and α-melanotropin. Cells treated with tunicamycin to block glycosylation produce very little of the biologically active peptides suggesting that the *N*-glycans control the accessibility of sites for proteolysis. Only in the fully glycosylated precursor are the flanking sequences rich in basic aminoacids preferentially cleaved by the pituitary enzyme to promote the sequence of reactions producing biologically active fragments [54].

4.2 Lectins: carbohydrate-binding proteins

4.2.1 General properties

A *lectin* may be defined as a protein of non-immune origin that precipitates glycoproteins and agglutinates cells and whose activity is inhibited by carbohydrates. This definition implies multivalency, i.e. the protein has at least two carbohydrate-binding sites to allow crosslinking between glycoprotein molecules. The first lectins to be studied were obtained from plants. Paul Ehrlich used the lectins of castor bean *Ricinus communis* and contributed to an understanding of lectin activity [1] which will be discussed in full later. The lectin, concanavalin A, of jack bean *Canavalia ensiformis* was crystallized by Sumner in the 1919 and is among the best characterized of lectins. Table 4.1 is a short compendium of the structure and specificities of plant lectins [55]. More recently lectins have been purified from animal tissues, and research in this area has been extremely revealing about the functional roles of carbohydrate-containing macromolecules.

The biological roles of lectins in plants are still being clarified. Various suggestions have been made: First, lectins may play a role analogous to the immune system in immunocompetent organisms, e.g. in fixing pathogenic microorganisms and bringing about their destruction. Secondly there is good evidence that lectins of leguminous plants play a role in colonization of the plant roots by nitrogen-fixing microorganisms such as *Rhizobia*. Plant lectins may have additional or even primary roles in other events such as germination. Whatever their roles turn out to be, plant lectins are very useful tools to study the functions of glycoproteins in cells and the specificity of carbohydrate–protein interactions. Such interactions have biological significance in several well-defined systems described in later sections and purified plant lectins are useful models to study these interactions.

4.2.2 Specificity

The sugar specificity of the lectins listed in Table 4.1 is derived from the simple carbohydrates that can inhibit cell agglutination by the lectins. Usually a particular monosaccharide, a methyl glycoside (α or β) or a disaccharide, e.g. gal $\beta 1 \rightarrow 3$ galNAc for peanut agglutinin, will inhibit lectin activity if used at millimolar concentrations. The specificity of a lectin is significantly greater than implied by such studies however, and extended carbohydrate sequences are often discriminated, analogous to the *endo*-glycosidases (see Chapter 2) and glycosyl transferases. Another analogy can be made with the bacteriolytic enzyme lysozyme. Lysozyme hydrolyses chitin and oligomers of *N*-acetylglucosamine joined in $\beta 1 \rightarrow 4$ glycosidic linkage. The best fit of substrate into the active cleft of lysozyme requires six monosaccharide residues. Smaller oligomers bind lysozyme but with reduced affinity and higher polymers bind no better than the hexasaccharide. The wheat germ lectin (Table 4.1) shows a rather similar specificity. A trisaccharide composed of *N*-acetyl-

Table 4.1 Properties of some purified lectins

Source	*Common name*	*Trivial name*	*Mol. wt.*	*Specificity*
Arachis hypogaea	peanut	PNA	110 000	gal
Canavalia ensiformis	jack bean	Con A	102 000	man
Ricinus communis	castor bean	RCA I	120 000	gal, galNAc
Ricinus communis	castor bean	RCA II (ricin)	60,000	galNAc, gal
Triticum vulgaris	wheat germ	WGA	36 000	sialic acids, $(\text{glcNAc})_2$

glucosamine residues linked glycosidically by $\beta 1 \rightarrow 4$ bonds is the preferred receptor for the lectin. It also binds *N*-acetylneuraminic acid although with lower affinity than the chitobiose-type structure.

Similarly, concanavalin A binds better to mannose-containing sequences than to mannose or simple mannosides. X-ray analysis of the lectin shows a carbohydrate-binding site on each of four identical subunits accommodating one or two monosaccharides. The minimum requirements for binding are a D-manno- or D-gluco-pyranoside configuration at C(3), C(4) and C(6) and the α-anomer is preferred. Therefore, although the binding site is small, the discrimination among carbohydrate sequences of glycoproteins can be great. Since glucose exists rarely in glycoproteins as a stable feature and terminal *N*-acetylglucosamine groups are also rare, the most important sugars contributing to concanavalin A binding by glycoproteins are α-mannosyl residues. Many of these are terminal residues in oligomannosidic glycans. Alternatively non-terminal mannoses that are substituted by glycosidic linkage to C(2) are able to bind. The main requirement for good binding [56] is two α-mannose residues substituting another residue in a branched structure (Fig. 4.1). This can be α-mannose linked to a core region sequence or the man_5 and man_6 derivatives, e.g. of ovalbumin, and good binding is preserved even after elongation of the chains with other *N*-acetylglucosamine, galactose and sialic acid units. Substitution of C(4) of one of the α-mannose residues however, drastically reduces binding affinity. More surprisingly, substitution of a

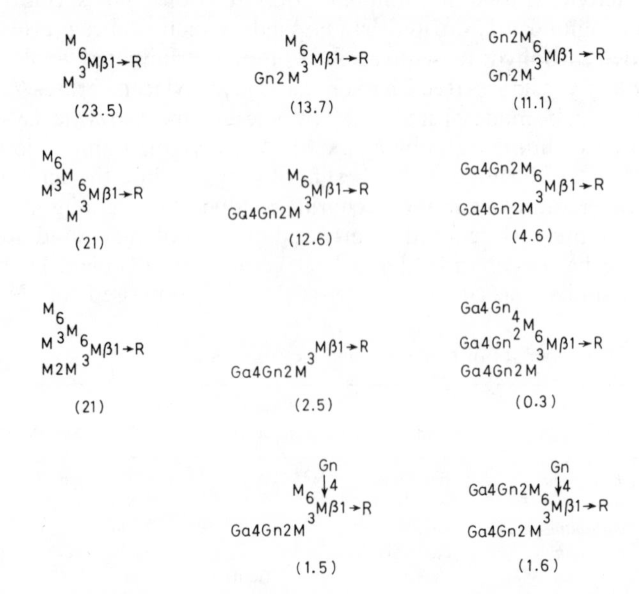

Fig. 4.1 Binding specificity of concanavalin A. Values show association constants ($\times 10^6$ M^{-1}) [56]. R, glcNAc $\beta 1 \rightarrow$ glcNAc . Asn.

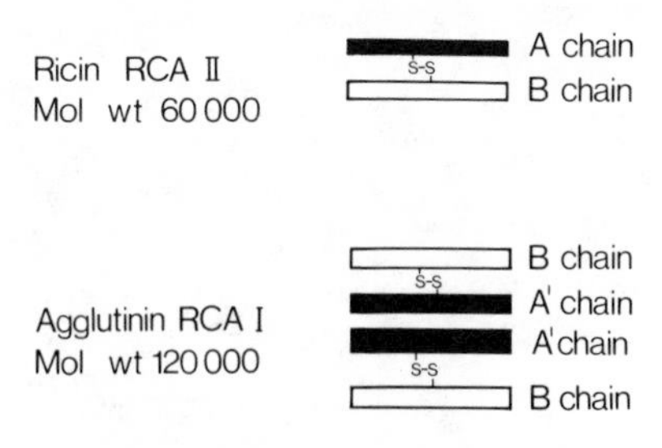

Fig. 4.2 Structure of castor bean lectins.

β-mannose residue with *N*-acetylglucosamine in the core sequence of

'hybrid' *N*-glycans also reduces binding to the outer branched α-mannose units. Presumably the *N*-acetylglucosamine residue prevents interaction with concanavalin A by steric shielding of the α-mannose units, or induces a conformational change (spreading) of these residues that is unfavourable for multivalent lectin binding.

Purification of glycoproteins and glycopeptides is often carried out by affinity chromatography on concanavalin A coupled to Sepharose or agarose. Only molecules containing carbohydrate binding sites with association constraints above about $4 \times 10^6\ s^{-1}$ are retained by the column allowing a simple separation of glycans of quite similar structure.

The castor bean lectins RCA I and RCA II (ricin) are multisubunit proteins of very similar structure (Fig. 4.2). RCA I agglutinates cells at low concentrations, has a molecular weight of 120 000 and is composed of two B chains each with a sugar binding site and two A^1 chains. Ricin agglutinates cells only at high concentrations and contains only one A chain similar to but not identical with A^1 and one B chain. The polypeptides of the two lectins are very similar in aminoacid composition and carbohydrate composition (the B subunit carries a 'high mannose' oligosaccharide) and *N*-terminal aminoacid sequence. Ricin however has a modified A subunit that is a powerful toxin [57]. The mechanism of ricin toxicity will be described later. RCA I and ricin differ significantly in their specificity of binding (Fig. 4.3). Generally RCA I binds to complex *N*-glycans with greater affinity than to *O*-glycans [58]. *N*-glycans carrying two terminal galactose residues are better at binding the lectin than glycans carrying a galactose sequence only on one branch. Unlike concanavalin A, RCA I and ricin bind very well to *N*-glycans in which an α-mannose residue is linked at C(2) and C(4) with galactose-containing sequences. Sialylation of *N*-glycan sequences does not affect appreciably lectin binding capacity but binding to *O*-glycans is completely abolished. The binding of the lectins to the galβ1 $\rightarrow$ 3 galNAc α1 $\rightarrow$ Ser (Thr) or galNAc α1 $\rightarrow$ Ser (Thr) sequences is greatly increased if many mono or disaccharides are distributed along a polypeptide backbone, as is the case in desialylated glycophorin for example.

	Sa6Ga4Gn2M $^{6}_{3}$Mβ→R Sa6Ga4Gn2M	Ga4Gn2M $^{6}_{3}$Mβ1→R Sa6Ga4Gn2M	Ga4Gn 4 Ga4Gn 2 M 6 $_{3}$Mβ1→R Ga4Gn2M
RCA I	7.6	15	15
RCA II	2.1	8.4	14.4

	Gn2M $^{6}_{3}$Mβ1→R Sa6Ga4Gn2M	Gn2M $^{6}_{3}$Mβ1→R Gn2M	Gn 4 Gn 2 M 6 $_{3}$Mβ1→R Gn2M
RCA I	3.4	0	0
RCA II	1.4	0	0

	Sa3Ga3Ganα1→R_2	Ga3Ganα1→R_2	Ganα1→R_2
RCA I	0	2.7	0
RCA II	0	7.7	0.7

Fig. 4.3 Binding specificities of castor bean lectins RCA I and RCA II (ricin). Values are association constants ($\times 10^6$ M^{-1}) [58]. R, glcNAc β1 → 4 glcNAc . Asn; R_2, polypeptide.

4.2.3 Toxic lectins

Ricin is one of the most toxic substances known. The Bulgarian broadcaster, George Markov, was shot on a London street in 1979 by an unknown person, probably for political reasons. A small hollow pellet was removed from Markov's body: from the dimensions only a couple of poisons would have been sufficient to kill a man. Ricin is one such substance and its presence was confirmed during post-mortem examination.

In less sombre vein ricin is a very useful tool to study the role of glycoproteins and the way by which small proteins such as hormones are taken up into cells. Ricin binds to cell surface carbohydrates through the B subunit (Fig. 4.4). These receptors are galactose-containing sequences as described previously. Subsequently the lectin enters into the cytoplasm where the A and B subunits dissociate and the A subunit binds to ribosomes to inactivate protein synthesis. Only a few ricin molecules are needed in the cytoplasm to completely block protein synthesis by catalytic inactivation of ribosomal factor(s) by the A subunit. Several other toxins with a similar mode of action are known. Each toxin is composed of two dissimilar subunits: one serves to bind to cell surface receptors and the other to affect some vital metabolic process to kill

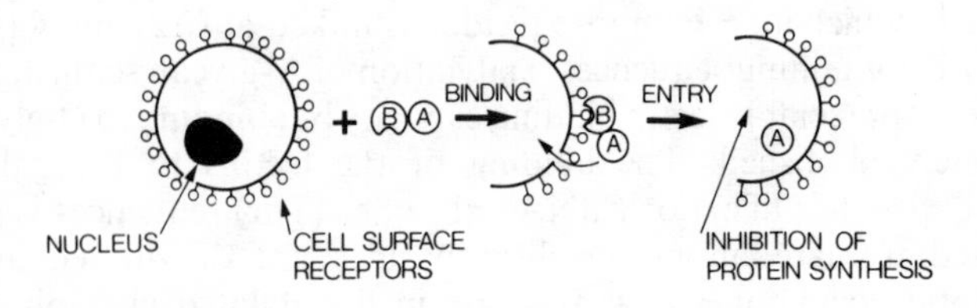

Fig. 4.4 Mechanism of ricin action.

sensitive cells. The toxic lectin of *Abrus precatorus* seeds, *abrin*, is very similar to ricin in sites of binding and inhibition of protein synthesis and *modeccin*, the lectin of *Adenia digitata* also binds to galactose receptors. The diphtheria toxin consists of a subunit with specificity for glycoproteins and a toxic subunit that effects ADP-ribosylation from NAD of elongation factor 2 (EF2) to inhibit protein synthesis. The exact sites of toxicity of other lectins are not as clearly understood but binding to carbohydrate receptors is an essential first step in each case and one factor in cytotoxicity may be to inhibit membrane transport of nutrients.

4.2.4 Lectin resistant cells

Cell lines resistant to the cytotoxic effects of lectins such as ricin can be isolated from mutagenized or unmutagenized stocks relatively easily. The resistant cell lines are very stable and fall into two main categories: cells that are unable to bind the lectins at the cell surface and cells that are unable to effect the entry of toxin once bound to the cell surface. The former class of mutants is especially valuable to study the functions of cell surface carbohydrates [59].

A common type of ricin-resistant mutant lacks the enzyme *N*-acetylglucosaminyl transferase I discussed in Chapter 3. Extracts of such mutants are unable to transfer *N*-acetylglucosamine from UDP-*N*-acetylglucosamine to a suitable acceptor such as ovalbumin glycopeptide V. Consequently *N*-glycan assembly is blocked at an early stage in these cells, and carbohydrate sequences build up that lack galactose-containing segments required for ricin-binding at the cell surface (Fig. 4.5). The carbohydrate sequences accumulating in the mutant cells are excellent receptors for concanavalin A and hence these cells show increased sensitivity to the weak cytotoxicity of this lectin.

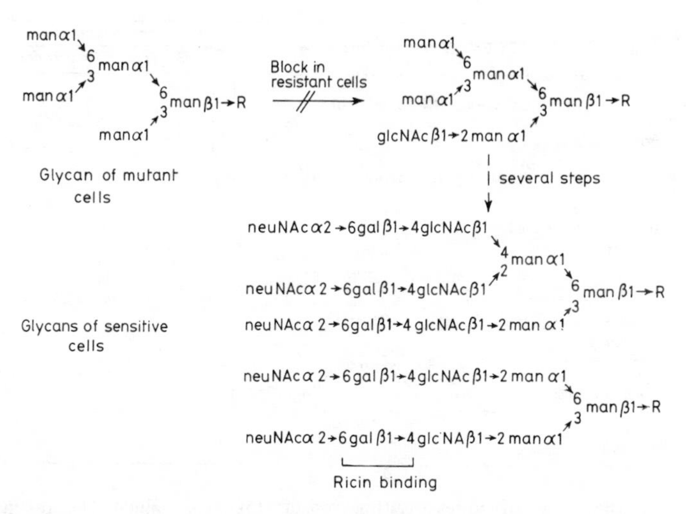

Fig. 4.5 Enzyme block in some ricin-resistant cells. R, glcNAc β1 → 4 glcNAc . Asn.

A very interesting change in N-glycan assembly is found in concanavalin A resistant cell lines (Fig. 4.6). Double mutants resistant to concanavalin A can be selected from glcNAc transferase I deficient mutants. These double mutants bind concanavalin A poorly due to the loss of one α-mannosyl residue (Fig. 4.6) as predicted from the binding specificity of the lectin (Fig. 4.1). This may be due to a deficiency of an $\alpha 1 \rightarrow 6$ mannosyl transferase involved in the assembly of an oligosaccharide–lipid intermediate: a $(\text{man})_7$ structure accumulates that is glycosylated, transferred to proteins and, in the absence of glcNAc transferase I, is processed to the structure shown [60]. Another type of mutation leading to concanavalin A resistance has already been mentioned. Cells lacking the ability to synthesize dolichyl mannose phosphate [27, 61] are resistant and show a reduced ability to bind the lectin. Evidently the glycoproteins made in this type of mutant carry fewer oligomannosidic sequences capable of binding concanavalin A with high efficiency.

4.3 Intracellular transport of glycoproteins

4.3.1 Effects of blocks in glycosylation

As described in Section 3.5, glycosylation is an early event in the passage of many proteins synthesized in the cisternae of the rough endoplasmic reticulum to their final cellular locations or secretion from the cell. One of the earliest functions ascribed to protein glycosylation was to facilitate secretion. Later refinements of this simple hypothesis propose that glycosylation is principally involved in the insertion of proteins into

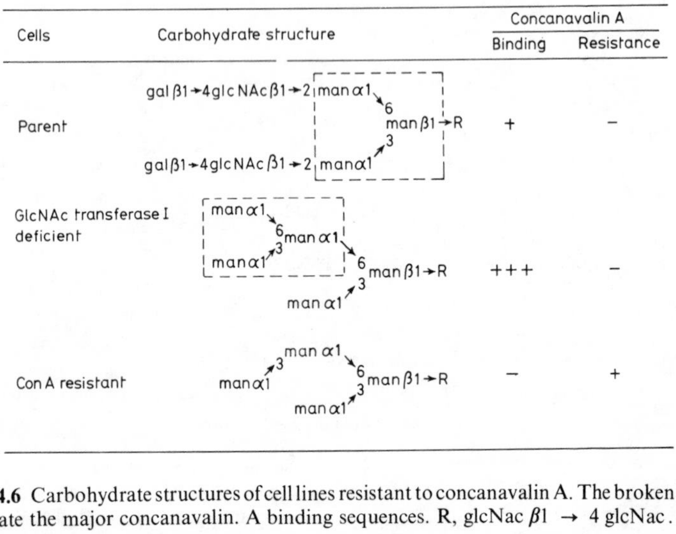

Cells	Carbohydrate structure	Concanavalin A	
		Binding	Resistance
Parent	galβ1→4glcNAcβ1→2 manα1 ↘6 man β1→R galβ1→4glcNAcβ1→2 manα1 ↗3	+	−
GlcNAc transferase I deficient	manα1 ↘6 man α1 ↘6 manα1 ↗3 man β1→R man α1 ↗3	+++	−
Con A resistant	manα1 →3 man α1 ↘6 man β1→R man α1 ↗3	−	+

Fig. 4.6 Carbohydrate structures of cell lines resistant to concanavalin A. The broken lines indicate the major concanavalin. A binding sequences. R, glcNac β1 → 4 glcNac . Asn.

membranes and their transport by membrane flow to other intracellular membranes and to the surface. The experimental evidence for this model is however very contradictory. Many proteins quite unequivocally do not need to be glycosylated for efficient intracellular transport from sites of polypeptide assembly to their final destination, as shown using drugs such as tunicamycin or glycosylation-deficient cell mutants.

A striking example of a rather selective effect of a block in glycosylation is in lymphoma cell lines resistant to concanavalin A and deficient in the synthesis of dolichyl mannose phosphate. These cells do not express at the surface a major glycoprotein antigen called Thy-1, although other glycoprotein antigens are found in normal amounts at the cell surface [62]. Possibly some proteins in the unglycosylated or partially glycosylated form are particularly sensitive to intracellular proteolysis whereas others are relatively resistant and survive intracellular passage more or less intact. A commonly observed feature of tunicamycin-treated cells is a highly distended endoplasmic reticulum engorged with unglycosylated proteins that normally would be swept into other cellular sites. A dramatic instance of such a phenomenon is in the human pathological condition known as α_1*-antitrypsin deficiency*. These patients synthesize in the liver a protein (protease inhibitor) that is similar to that of normal individuals which is secreted into the blood stream. Interestingly the *N*-glycans attached to the protein in pathological liver appear to be partially processed oligomannosidic chains unlike the complex, sialic acid containing glycans of the product secreted by normal individuals [63]. Of course, it is difficult to decide if a normal intracellular route is blocked *because* of some defect in *N*-glycan assembly or if the immature glycosylation simply reflects the stage at which the intracellular transport is stalled due to other unknown causes.

4.3.2 The life cycle of lysosomal enzymes

Over the last five years research on the intracellular cycling of one particular group of glycoproteins has provided direct evidence for a critical role of normal glycosylation. These glycoproteins are the collection of hydrolases present within lysosomes. A unique carbohydrate-based recognition system has been discovered that has general significance as well as pin-pointing a biochemical defect responsible for a human disease state.

The rare human condition is known as I-cell disease, not to be confused with the Ii antigens; the nomenclature is unfortunate but is well-established in the literature and cannot be easily changed. The affected individuals die at an early age and show a massive accumulation within the cells of high molecular weight glycoconjugates. The basic defect is of a normal complement of intracellular lysosomal hydrolases responsible for the degradation of ingested connective tissue components. Cells such as skin fibroblasts cultured from affected individuals show a similar abnormality and also secrete abnormally high levels of lysosomal hydrolases into the culture medium. In other words, the cells secrete hydrolases that in the normal condition are sequestered within lysosomal vesicles to accomplish the degradation of engulfed macromolecules and cell debris [64]. There is a receptor on the surface of normal and I-cells that recognizes some structural feature of lysosomal hydrolases and triggers the adsorptive endocytosis of the hydrolases. The I-cells are quite capable of mediating uptake of hydrolases obtained from normal cells but are unable to recognize the enzymes synthesized and secreted by themselves. Furthermore, normal cells are also unable to recognize the I-cell enzymes. Clearly therefore, the defect is in the structure of the I-cell hydrolases and not in a block in adsorptive endocytosis or receptor function of the cells. A role for the carbohydrate component of lysosomal hydrolases was first established by the finding that mannose-6-phosphate competitively inhibits the uptake of normal hydrolases by normal fibroblasts, demonstration of the presence of phosphorylated *N*-glycans in a variety of lysosomal enzymes, and the conversion of these enzymes to a low-uptake form by treatment with phosphatases. The enzymes secreted by I-cells lack this carbohydrate structure, do not bind to the surface of either normal or I-cells and are not taken into the cells. Although the mannose-6-phosphate receptor is present in small amount at the cell surface and was first recognized there, it now seems that at least 90% of the receptors are present in the endoplasmic reticulum, Golgi membranes and the lysosomes. Based on this evidence the receptors would appear to be involved primarily in the delivery of newly synthesized hydrolases from the rough endoplasmic reticulum into smooth membranes and eventually into lysosomes (Fig. 4.7).

The following sequence of events can be proposed. After glycosylation of the nascent polypeptide with the oligomannosidic intermediate, phosphorylation of selected mannose units occurs. The phosphorylation is not a random event: certain mannose residues are especially favoured (Fig. 4.8, $^{++++}$) and some are never phosphorylated.

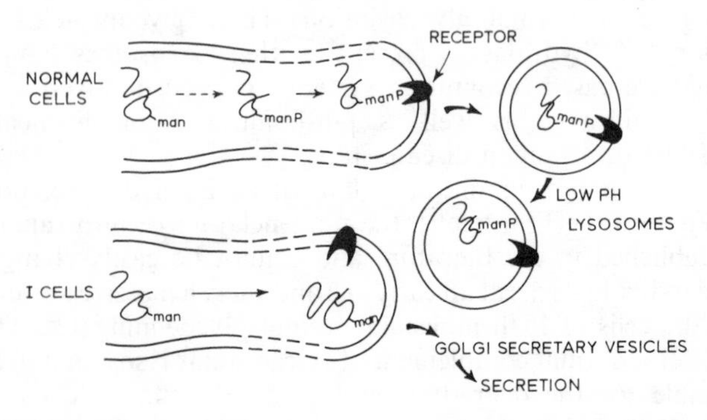

Fig. 4.7 A scheme for the intracellular transport of lysosomal hydrolases and the defect in I-cells.

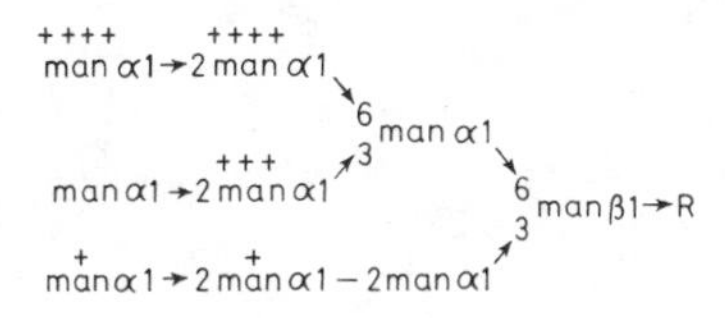

Fig. 4.8 Phosphorylated *N*-glycans of lysosomal enzymes [65]. The degree of phosphorylation of individual mannose residues is indicated by plus signs.

In many cases a single *N*-glycan may have two phosphorylated mannose units and a whole family of species apears to be present in a single enzyme such as *β*-glucuronidase. The reason for such heterogeneity remains unknown [65]. The mannose-6-phosphate units are recognized by specific receptors located on the cisternal side of regions of the ER–Golgi elements that eventually bud off into primary lysosomes containing a full complement of hydrolytic enzymes held at the intravesicular side of the lysosomal membrane. At the low pH characteristic of lysosomes, probably due to a very high concentration of sialic acid-containing glycoproteins exposed at the intra-vesicular surface, the binding between the mannose-6-phosphate groupings and the specific receptors is weakened and the hydrolases become more or less freely dissolved in the lysosomal sap. At this stage the glycans are dephosphorylated to remove the recognition signal [66]. Macromolecules and cell debris entering the cells by adsorptive endocytosis in the form of endocytotic vesicles fuse with the primary lysosomes, become exposed to the full complement of lysosomal enzymes at their optimum pH of 4.5 and are rapidly degraded. In the I-cells phosphorylation of the *N*-glycans appears to be blocked, the recognition signal for sequestration in regions of the ER–Golgi elements destined to form lysosomes cannot then take place and the lysosomal enzymes are redirected into secretory vesicles.

Recent experiments have modified this simple scheme somewhat [67, 68]. The phosphorylation of mannose residues of the lysosomal hydrolases does not take place directly, for example by kinase action, from ATP. Rather the initial reaction is transfer of α-*N*-acetylglucosamine-1-phosphate from UDP-*N*-acetylglucosamine to the C(6) of a mannose residue, probably after transfer and partial processing of the usual oligogluco-mannosidic unit from the lipid intermediate to the protein (Fig. 4.9). Since this reaction is very selective and takes place presumably only with lysosomal enzymes it most likely occurs in the later stage of differentiation of the endoplasmic reticulum into lysosomal precursors or the enzyme recognition site includes a domain, e.g. a peptide segment that is present exclusively in lysosomal enzymes and would exclude other glycoproteins with 'high mannose' glycans. Possibly, for example, the specific transferase is located close to the

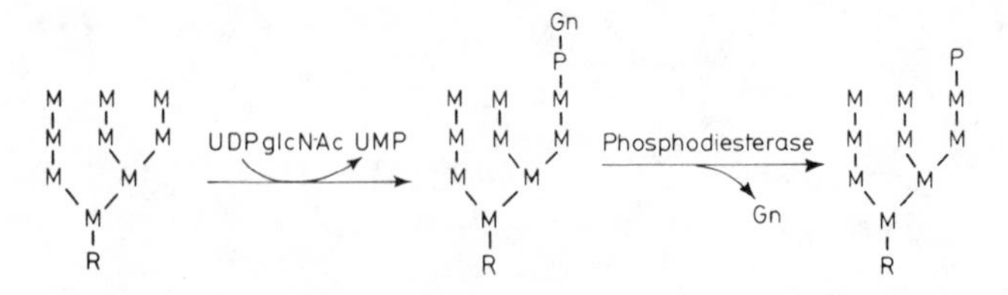

Fig. 4.9 Biosynthetic intermediates of lysosomal β-glucuronidase. R = the core region of *N*-glycans. Only one phosphorylated form is shown. See Fig. 4.8 for other isomers.

mannose-6-phosphate receptors on the cisternal side of specialized regions of the endoplasmic reticulum. In the second enzymic step the mannose-6-phosphate signal is generated by a highly specific α-*N*-acetylglucosaminyl phosphodiesterase which releases free *N*-acetylglucosamine and triggers binding of the lysosomal enzymes to the receptors. The predominant site for both enzymes is in purified, smooth, Golgi-derived membranes, which is consistent with the basic proposal. In agreement with the demonstration of the biochemical defect in I-cell as a failure to produce a stable phosphorylated form of the lysosomal enzymes, the specific *N*-acetyl glucosamine-1-phosphate transfer reaction is totally lacking in these cells.

The phosphomannosyl-enzyme receptor is widely distributed as expected since lysosomes are a common feature of cells.

4.4 Catabolism and clearance

Proteins within cells are degraded in a controlled manner during normal growth and differentiation [69]. The degradation and turnover of proteins is determined in part by inherent features, e.g. conformation of the polypeptide, but there is considerable evidence that the carbohydrate moieties also regulate the catabolism of glycoproteins. For instance in the liver, plasma membrane glycoproteins turn over more rapidly than non-glycosylated proteins and in cultured cell lines the plasma membrane glycoproteins are replaced almost entirely under non-growing conditions within a few hours [18].

A carbohydrate-based signal for glycoprotein catabolism has been discovered which may be relevant to these processes. The glycoproteins are those present in the blood stream which are taken up by the liver for catabolism.

4.4.1 Soluble glycoproteins

The system to be described was discovered by chance. Gilbert Ashwell and Anatole Morell and their colleagues wished to follow the fate of the glycoprotein ceruloplasmin when injected into the blood stream. To do this the glycoprotein was treated with neuraminidase to expose galactose residues present on the *N*-glycans of the glycoprotein (see Fig. 2.5) which could then be radioactively labelled by treatment with galactose oxidase to generate an aldehyde at C(6) followed by reduction with tritiated sodium borohydride. Surprisingly, the labelled glycoprotein disappeared within a few minutes from the blood stream and appeared in the liver of rats whereas the unmodified glycoprotein survived for days. Subsequent analysis of this interesting phenomenon has revealed a lectin-like activity on the surface of hepatocytes which binds terminal galactose residues of a variety of asialoglycoproteins and mediates uptake of galactoproteins into lysosomes for hydrolysis by the acid hydrolases [70]. This process therefore provides an explanation for the early finding that orosomucoid of human blood is almost entirely sialylated: very few, if any, galactose residues are exposed.

The galactose binding lectin is a multimeric (probably dimeric) glycoprotein firmly embedded into hepatic membranes and solubilized only by treatment with detergents. The subunit molecular weight of about 45 000 distinguishes it from other galactose-specific lectins probably involved in intercellular adhesion (see Section 4.5.1). The hepatocyte lectin agglutinates neuraminidase-treated cells and is mitogenic for lymphocytes, i.e. stimulates resting blood lymphocytes into DNA synthesis. Both of these properties are diagnostic of multi-valency of binding sites on one functional lectin molecule. The binding and uptake of glycoproteins into hepatocytes also relies on multiple interactions with the lectin [71]. Glycoproteins or purified glycopeptides with multiple terminal galactose residues are bound with high affinity and are rapidly taken up into the cell. Glycoprotein with complex glycans containing just two terminal residues bind poorly and do not trigger endocytosis (Fig. 4.10). Thus asialo-orosomucoid with four outer branches in the *N*-glycans (Fig. 2.11) is an excellent substrate for

		Binding affinity	Rate of endocytosis
(a)	asialo-orosomucoid	High	High
(b)	Ga–Gn–M(Ga–Gn)–M–$(Gn)_2$–R, with M–Ga–Gn (three Ga–Gn branches)	High	High
(c)	Ga–Gn–M and Ga–Gn–M on M–$(Gn)_2$–R	Low	Low
(d)	Ser.Pro.Pro.Gly.Asp.Asp.Thr. with $(Ga)_{0 \text{ or } 1}$–Galn on Ser and $(Ga)_{0 \text{ or } 1}$–GalN on Thr	Low	Low
(e)	Ser.Thr.Pro.Pro.Thr.Pro.Ser.Pro.Ser. with $(Ga)_{0 \text{ or } 1}$–GalN on the last two Ser	High	High

Fig. 4.10 Binding and endocytosis of glycoproteins by hepatocytes. *O*-glycans containing terminal galactose residues behave similarly to the same glycopeptides from which galactose is removed. R, asparagine residues.

hepatic endocytosis while serum asialo transferrin with just two branches is a very poor substrate [72]. Also of interest is the very high affinity of binding and efficient endocytosis of *N*-acetylgalactosamine-based *O*-glycans (Fig. 4.10). Like the requirement for a multiplicity of closely spaced galactose-terminated branches of *N*-glycans, glycopeptides carrying closely-spaced *N*-acetylgalactosamine-linked glycans are much more efficient at binding to hepatocytes and triggering uptake. Furthermore the *N*-acetylgalactosamine terminal residues do not have to be substituted with galactose to bind the receptor.

The high affinity of certain glycopeptides for hepatocytes can be used to direct proteins from the circulation into the liver. For example, the glycopeptide (b) (Fig. 4.10) coupled to albumin induces the rapid clearance of this normally non-glycosylated and stable blood component [70].

Binding of glycoproteins to hepatocytes is absolutely dependent on Ca^{2+} ions and is completely inhibited by treatment of hepatocytes or the purified lectin with neuraminidase. This is because removal of sialic acid from the glycan chains of the lectin induces self-aggregation through interaction with exposed galactose terminal residues.

The lectin seems to be confined to liver where it is confined to the face of the hepatocyte in direct contact with the blood flow. This specific localization would argue for a primary role of the lectin in the clearance of partially degraded or modified glycoproteins from the serum or possibly a role in the intracellular transport of serum glycoproteins made in the liver into the blood stream.

A direct demonstration of the *in vivo* function of the hepatocyte lectin in clearance comes from two sets of experiments: (1) the simultaneous injection of asialoglycoproteins into the blood stream with antibodies directed against the purified receptor significantly prolongs the time in the circulation of those glycoproteins [73]; (2) higher than normal amounts of galactose-terminal glycoproteins appear in the blood of patients with liver cirrhosis. However, only about 10% of receptors are expressed at the cell surface at any one time: the majority of receptors are bound to smooth membrane and Golgi membranes that may in part represent precursors of vesicles carrying receptors to the cell surface thus continually replenishing surface receptors lost by the endocytotic events [74]. It has been calculated for example that the entire population of surface receptors is removed into the cell every 5–10 minutes or so and a large intracellular pool is obviously a necessary feature of the clearance process. The replenishment of surface receptors also appears to be assisted by a recycling of receptors from lysosomes and this implies that the receptors are protected from lysosomal degradation. A mechanism for protection of the active site of the receptors is suggested by the surprising finding that these sites face the cytostolic space in the lysosomes. Hence there is an apparent inversion of the receptor molecules in the membrane after lysosomal fusion. If these receptors are to appear in the correct orientation again at the cell surface there presumably is a second inversion of the receptors in the membrane system shuffling them back to the cell surface. The mechanisms involved are unknown at present.

In the normal mammalian liver, glycoproteins circulating in the blood stream are exposed to the surface membranes of non-parenchymal cells lining sinusoids as well as the hepatocytes. The non-parenchymal cells include Kupffer cells and endothelial cells and the former cells in particular also play an important role in glycoprotein clearance. If the glycoprotein orosomucoid is treated with neuraminidase and then β-galactosidase to expose terminal *N*-acetylglucosamine the modified glycoprotein is rapidly taken up by the reticuloendothelial system (Fig. 4.11). Mannose-terminated glycoproteins are similarly cleared: for example, a variety of rat lysosomal enzymes including β-glucuronidase and β-*N*-acetylglucosaminidase are also cleared rapidly and the simultaneous infusion of asialo-agalacto-orosomucoid blocks their clearance [75]. It will be recalled that the lysosomal enzyme have *N*-glycans rich in oligomannosidic chains. This experiment suggests therefore, that the recognition molecule on the surface of Kupffer cells binds to glycoproteins carrying either *N*-acetylglucosamine or mannose terminal residues. The lectin is widely distributed on professional phagocytic cells, for example alveolar macrophages, and is a second major carbohydrate-based catabolic mechanism in mammalian tissues. *N*-acetylglucosamine mannose-binding components which may be the surface lectin of non-parenchymal cells have been isolated. A multimeric membrane-bound glycoprotein of rat liver has a subunit molecular weight of 32 000. It is antigenically unrelated to the hepatocyte lectin but similarly requires Ca^{2+} ions for activity. Glycoproteins terminating in mannose residues are bound slightly better than those with terminal *N*-acetylglucosamine residues and the specificity of the lectin is similar therefore to con-

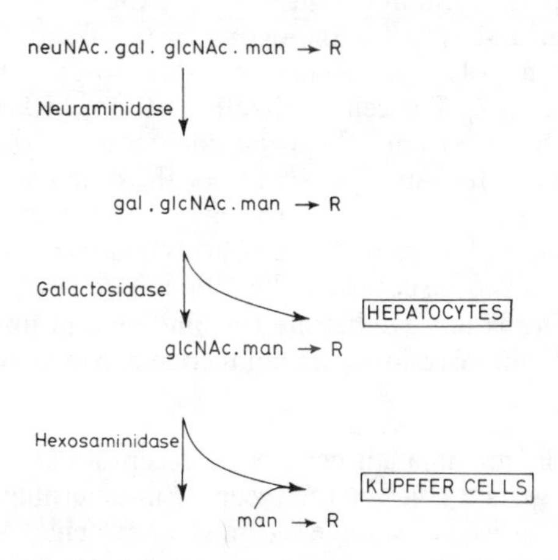

Fig. 4.11 Uptake of modified glycoproteins by the mammalian liver. R, core region of *N*-glycans.

canavalin A [76]. It will be recalled that concanavalin A binds to internal mannose residues even when these are substituted with other sugars, e.g. *N*-acetylglucosamine on C(2) although the affinity of binding is greater for glycoproteins with unsubstituted mannose terminals (see Fig. 4.1).

The functions of the hepatocyte and Kupffer cell lectins may well include the clearance of partially degraded serum glycoproteins or lysosomal hydrolases from the blood stream. The latter function would seem to be particularly important in preventing damage to the surface of cells exposed to the circulation. Yet another function may be to facilitate the destruction of microorganisms such as yeasts with a high complement of oligomannosidic chains. The ingestion of yeasts by alveolar macrophages, for example, is inhibited by mannose and glycoproteins with terminal mannose residues [77].

Unlike mammalian species the blood glycoproteins of birds have many exposed galactose residues. Clearly, therefore, a system different to the mammalian hepatocyte receptor would be expected. The terminal residues recognized are *N*-acetylglucosamine and treatment of galactose-terminal glycoproteins with *β*-galactosidase causes the rapid clearance of these modified glycoproteins from the blood stream. The specific lectin has been isolated from chicken liver. It is a glycoprotein of subunit molecular weight 26 000. The complete aminoacid sequence has been completed and shows several interesting features. The acetylated *N*-terminal region is followed by a peptide segment enriched in aminoacids with non-charged, largely hydrophobic side chains (residues 25–48) which is the probable site for integration of the lectin into the cell membrane. The carboxyl terminal region probably extends into the extracellular space and contains the single *N*-glycan on asparagine residue 67. The seven cysteines are also present in this long segment which is probably therefore highly folded. Evidently the orientation of the chicken hepatic lectin in the membrane is the reverse of that found for glycophorin and is more analogous to the Band 3 glycoprotein except the lectin polypeptide chain passes through the membrane interior just once [78]. The cell localization of the chicken lectin is not known as yet but it is unlikely to be analogous to the Kupffer-cell receptor of mammalian species because of the difference in molecular weight, cation requirements and carbohydrate specificity. Thus, after removal of sialic acid, galactose and *N*-acetylglucosamine from orosomucoid, the modified glycoprotein bearing a high content of terminal mannose residues is not a substrate for binding and uptake, showing that mannose-terminal residues are not recognized by the chicken lectin.

4.4.2 Cells

The fact that the mammalian hepatocyte receptor can agglutinate red blood cells suggests a role for the receptor in determining the fate of circulating cells as well as soluble serum glycoproteins. As described in Chapter 2, the human red cell membrane contains a major highly sialylated glycoprotein, glycophorin, and analogous components are found on the erythrocyte surface of most species. When such cells marked with a radioactive label are treated with neuraminidase and infused into the blood stream of rats, rabbits or dogs the cell-associated radioactivity disappears rapidly from the blood and is sequestered in the liver and spleen, whereas the untreated cells are much more stable [79]. However, all the evidence argues against the same hepatocyte receptor mediating the clearance of soluble glycoproteins and erythrocytes. Firstly, treatment of desialylated glycoproteins such as ceruloplasmin with galactose oxidase to convert the C(6) of terminal residues to an aldehyde group provides protection on the glycoprotein. The modified glycoproteins persist in the blood almost as long as the control glycoprotein complete with sialic acid. It appears therefore that the hepatocyte receptor for asialoglycoproteins requires a primary C(6) hydroxyl group. By contrast galactose oxidase treatment of desialylated erythrocytes does not prolong their short-lived circulation. Secondly, desialylated red cells do not bind to isolated hepatocytes but adhere strongly to Kupffer cells and mononucleated cells of spleen. The binding to macrophages is inhibited by galactose suggesting as one possibility yet another carbohydrate-based recognition system for the clearance of desialylated erythrocytes distinct from the hepatocyte receptor for galactose-terminal glycoproteins [80].

These interesting sets of data may ultimately be explained by the respective binding specificities of the galactose inhibitible lectins of hepatocytes and Kupffer cells. The major sialylated glycoprotein of the human erythrocyte membrane, glycophorin, for example, differs from most serum glycoproteins in the presence of an *N*-acetylglucosamine linked to the central β-mannose unit of the core region. Perhaps this feature dictates the uptake of desialylated erythrocytes into Kupffer cells rather than hepatocytes although terminal galactose is the dominant sugar recognized by both systems [81].

Untreated erythrocytes isolated from a normal blood sample are slowly removed from circulation. Changes in surface carbohydrate structures have been detected in red cells of different ages. Older cells contain 10–15% less sialic acid than newly formed cells. There is a normal replenishment of red cells occurring constantly and it is possible that the experiments just described reflect a mechanism by which the life of a red cell in the blood stream is regulated.

4.5 Cell adhesion

4.5.1 Cell–cell adhesion

Cell adhesion is a property of fundamental importance in the formation of tissues and organs during embryogenesis. Embryonic development requires differentiation of totipotent cells, cell sorting into organized masses of similar cells, cellular translocations and terminal development into specific organs. These processes depend upon specific intercellular recognition, and selective adhesion between cells. Classical experiments with cells dissociated from embryos or differentiated tissues such as heart, kidney, liver and so on, tell us that adhesive specificity is maintained by differentiated cells. Cells dissociated from various organs and mixed in suspension tend to re-form aggregates of cells in a *tissue-specific* manner; cells derived from heart tissue exclude liver-derived cells during aggregation, for example.

As mentioned previously, interference with normal glycosylation often prevents development, for example in sea urchin embryos as well as in some developing mammalian systems, e.g. kidney. Hence glycoproteins are assumed to play important roles in many of these adhesive interactions leading to normal development.

It is reasonable to assume that adhesive interactions involve binding between complementary molecules. If the carbohydrate moieties of glycoproteins are involved in cell adhesion two types of complementary interactions are possible: (1) between carbohydrate chains by a process analogous for example to the chain–chain interactions of many linear polysaccharides [4]; (2) between a carbohydrate moiety and a carbohydrate binding protein, e.g. a lectin. The second model (Fig. 4.12) has gained pre-eminence in recent explanations because direct interactions between glycans of the type existing in glycoproteins have not yet been shown to occur and because carbohydrate-binding proteins of the required specificity and cell-surface location have been identified. In some cases these proteins appear at a time when developmental events take place, strengthening the hypothesis that protein–carbohydrate interactions are involved. The interactions may be between molecules integrated into the cell surface membrane of adherent cells (Fig. 4.12a) or may be mediated by aggregation factors (Fig. 4.12b, c). In the latter case the aggregation factors may be products secreted into the intercellular space by either interacting cell or by some cell remote from the site of specific cell–cell adhesion. This cell would then have an inductive effect on the aggregation of other cells, a phenomenon very often encountered in differentiation.

Aggregation factors have been isolated from several tissues including sea-water sponges, historically one of the first organisms to be studied for specific adhesive mechanisms. Early in this century Edward Wilson showed that sponges dissociated into single cells when placed in fresh water and would aggregate if resuspended in sea water. The effect is regulated by the salt constituents in particular calcium of sea water. If

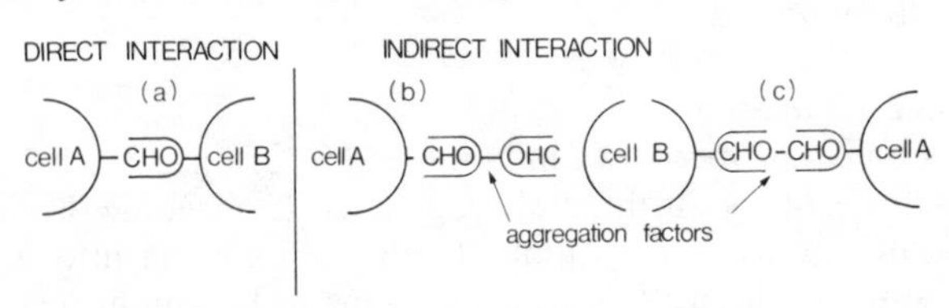

Fig. 4.12 Carbohydrate–protein interactions in cell–cell adhesion. CHO, carbohydrate moiety; ⊱ , a carbohydrate-binding moiety.

two different sponges are dissociated, mixed and incubated in sea water, cells derived from each sponge form aggregates with exclusion of the foreign sponge cells. This phenomenon can be very easily seen if sponges of different colours (red or green due to their content of endogenous chromophores) are mixed. Red or green aggregates are formed and not mixed aggregates (Fig. 4.13).

The aggregation factors released from sponge cells upon dissociation can be readily separated from the cells by centrifugation and are high molecular weight glycoproteins of unusual composition [82]. In addition to galactose, mannose and *N*-acetylhexosamine, sulphate and uronic acids are present and the latter is clearly an important constituent since addition of free glucuronic acid inhibits aggregation of the sponge cells in the presence of the aggregation factor. However the aggregation factors show the specificity required: the factor from the sponge *Microciona parthena* for example cannot induce aggregation of unrelated species of sponges such as *Haliclona*.

A similar system has been identified in another sponge *Geodia cyndonium*. Interestingly, the partially purified aggregation factor is associated with a glucuronyl transferase activity suggesting that a conversion of a non-aggregating system to an adhesion-competent one may occur by an extracellular addition of glucuronic acid residues provided, of course, that these sites are provided with the necessary activated sugar derivate [83].

Similar aggregation factors have been isolated from the extracellular fluids of cultured cells of higher organisms, e.g. primary cultures of embryonic chick neural retina [84]. This factor aggregates retina cells but not cells from other tissues. It is a glycoprotein containing monosaccharides characteristic of *N*-glycans. Another cell aggregating factor is secreted by mouse ascites teratoma cells in culture and its aggregating activity is inhibited by β-galactosidase treatment or by addition of free galactose [85]. Evidently galactose residues present in the aggregation factor are important in mediating cell–cell interactions in this system (Fig. 4.12c).

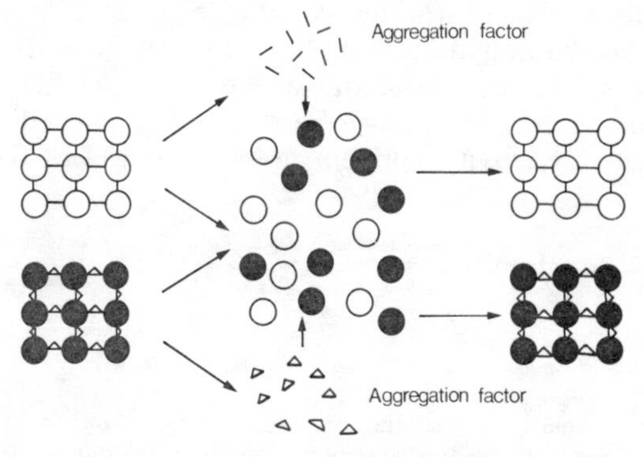

Fig. 4.13 Dissociation and specific re-aggregation of cells.

Other evidence for an involvement of glycoproteins and galactose residues in particular in cell–cell adhesion is indirect but nonetheless very suggestive. Single cell suspensions of baby hamster kidney (BHK) cells will readily form aggregates in calcium-containing media, a process stimulated by prior treatment of the cells with neuraminidase and blocked by treatment with galactose oxidase to convert the C(6) hydroxyl group of galactose to an aldehyde. Similar results have been obtained with other cells and in one case β-galactosidase caused a significantly decreased rate of aggregation. These results support a role in adhesion for complex *N*-glycans having the terminal sequence neuNAc. gal. glcNAc → R, where R is the core region sequence (Fig. 4.14). Further indirect evidence comes from a study of the adhesive properties or ricin-resistant BHK cell mutants described earlier (Section 4.2.4). Many of the resistant cell lines aggregate poorly and a linkage of this phenomenon to a block in glycosylation and a decrease in ricin-binding surface carbohydrates can be demonstrated. Finally, cells attach to inert agarose beads provided galactose is covalently linked to the beads (other sugars cannot be substituted for galactose) [86].

This evidence taken together is suggestive of a mechanism involving direct interactions (Fig. 4.12a) between galactose-rich carbohydrate chains and a carbohydrate binding protein on the surface of apposed cells (Fig. 4.14). Saul Roseman made the first suggestion for the nature of the carbohydrate-binding proteins involved in such adhesive interactions [87]. He pointed out the extreme specificity of glycosyl transferases, which therefore appear as attractive candidates for mediating adhesive interactions of high specificity. Since it is unlikely under normal circumstances that activated sugar precursors are present in the intercellular space, glycosyl transferases, if located at the cell surface, could participate in stable interactions with appropriate carbohydrate sequences that function as acceptors in the enzymatic reactions catalysed by the enzymes. The model is attractive because it suggests a way in which adhesion between cells might be dissociated in a controlled manner. If activated sugars became available at the contact sites between cells then sugar transfer would take place with dissociation of the transferase–carbohydrate complex. The early evidence for surface-located glycosyl transferases is equivocal however and has been heavily criticized. However, some recent experiments do support the notion that a sialyl transferase activity is present at some cell surfaces. It still remains

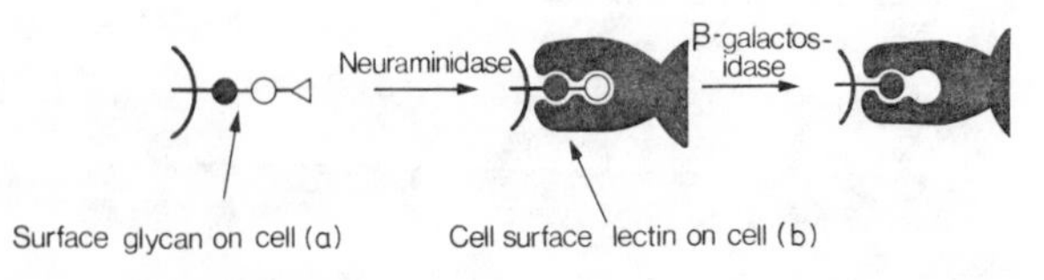

Fig. 4.14 Adhesion mediated by cell surface carbohydrate and protein receptors. Adhesion is enhanced by neuraminidase treatment of cells to expose additional terminal galactose residues. β-Galactosidase treatment reduces adhesiveness.

to be established, however, that these transferases participate in adhesive interactions. Another possibility is that glycosidases located at the cell surface may play a role by interacting non-hydrolytically with carbohydrate sequences on apposed cells. Lysosomal enzymes which have rather low pH optima may well form such stable binding interactions at the intercellular pH which is likely to be considerably higher than the intra-lysosomal pH required for optimal catalytic activity. Direct evidence in support of such a theory comes from the localization of small amounts of lysosomal enzymes at cell surfaces and the fact that cells will stick to inert glass surfaces to which neuraminidase, β-galactosidase and α-mannosidase have been attached. These experiments certainly need further consideration [88].

Currently however the most popular candidates for the carbohydrate-binding molecules are lectins with no known enzymic activity. A number of such lectins have been isolated. For example, in *Dictyostelium discoideum* a lectin appears when the unicellular amoeboid form of the slime mold is starved of food. The cells then become able to form aggregates that differentiate into a multicellular organism composed of a stalk and a fruiting body consisting of many spores. The lectin, as isolated, is inhibited by *N*-acetylgalactosamine, galactose and D-fucose, i.e. sugars of the D-galacto series. It now appears that the slime mold contains at least two such lectins called *discoidins* (I and II). Similar lectins (purpurins, pallidins) are known in other slime molds. *Dictyostelium purpureum* and *Polysphondylium pallidum*. Discoidin, purpurin and palladin lectins show some specificity towards the homologous slime mold. Presumably different carbohydrate sequences unique to each slime mold are involved in aggregation and the different lectins discriminate between these, although simple galactosides at high concentrations can inhibit cellular aggregation showing that this sugar is the dominant determinant in each case. Evidently discoidin I and II and the purpurins appear at different times in development suggesting that there are small changes in surface carbohydrates accompanying development, each recognized by the appropriate lectin appearing in a regulated manner at the cell surface. Probably each lectin is important for normal development, since a mutant of *Dictyostelium discoideum* which is unable to aggregate normally lacks functional discoidin I molecules but does contain surface receptors for discoidins isolated from the normal organism. The slime mold lectins are oligomeric proteins with subunits of about 25 000 molecular weight [89].

Another class of lectins has been isolated from a variety of avian and mammalian tissues. These lectins are characterized by a lower molecular weight approximately 13 000, a cell agglutinating activity that is inhibited by galactosides and possession of free sulphydryl groups. In the absence of reducing agents the lectin subunits aggregate, probably by disulphide bond formation and become inactive. The lectins are not integral constituents of the cellular membranes since they are solubilized by soaking the tissues in simple galactosides like lactose. Hence the lectins appear to be aggregation factors acting outside the cell to stick cells together by combining with appropriate integral membrane glycoproteins (Fig. 4.12b).

The role of these galactose-binding lectins in differentiation is indicated by several lines of evidence. The chick pectoral muscle lectin appears between 8 and 16 days of embryonic life during the period of active muscle differentiation. In a myoblast cell line (L6) lectin activity increases during fusion of the myoblasts into myotubes; cell–cell adhesion necessarily precedes fusion and the lectin is detected on the myoblast cell surface by immunofluorescence. Paradoxically however, most of the lectin is located intracellularly. Perhaps a large intracellular pool is necessary to ensure a continuing supply of surface-located lectin during the fusion process. Alternatively the lectin may serve several different functions, some of them concerned with the transport of glycoproteins within the cell. However the evidence does support a role for the lectin in specific interactions with surface galactose residues in the initial stages of muscle differentiation [89].

Chick embryonic skeletal muscle contains a second lectin which also changes in activity with muscle development. Its cell agglutinating activity is strongly inhibited by heparan and heparan sulphate, an important constituent of the extracellular matrix. The lectin is secreted into the extracellular space by differentiating muscle cultures and may be more important in the adhesion of muscle cells to a supporting matrix (see Section 4.5.2) rather than in mediating cell–cell adhesion [89].

An interesting system in which a galactose-specific lectin plays an important role is erythroid development. Erythroblasts display a typical 13 000 molecular weight lectin on the cell surface that clusters these cells in 'erythroblastic islands' surrounding a nursing macrophage. The concentration of the lectin at the cell surface decreases with maturation of the erythroblasts and release from the aggregates of reticulocytes, the direct precursors of erythrocytes [90]. There are extensive changes in galactose-containing carbohydrate sequences during erythroid differentiation at least in human cells as detected by reagents reacting with the Ii antigenic determinants (Chapter 3). Lectins similar to those present at the erythroblast surface are known to react with Ii determinants [91]. It will be interesting to see if any parallels can be established between the appearance of the galactose-binding lectin and modulation of Ii determinants in the developmental system.

Finally a galactose-binding lectin is released from neonatal rat brain with lactose and therefore is analogous to many of the galactose-binding lectins described here. The brain lectin appears in 10 day old rats at a time of maximum formation of synapses and perhaps the lectin plays an important role, e.g. synaptogenesis in the developing rat cortex [92].

4.5.2 Cell adhesion to substratum

All cells except freely circulating cells such as erythrocytes or peripheral lymphocytes are surrounded by or embedded in a layer of carbohydrate-rich material. This layer is sometimes called the *cell coat* or *glycocalyx* or *extracellular matrix*. Connective tissues such as cartilage bone or tendon are extreme examples of in some areas a virtually acellular highly differentiated extracellular matrix laid down by the specialized cell types within it or in close apposition to it. The extracellular matrix provides a protective environment excluding harmful substances from the cell surface, maintains a buffered layer to compensate for sudden shifts in the composition of the fluids bathing the cell, and acts as a relatively rigid structure supporting body shape. In addition, the extracellular matrix plays important roles in controlling cell growth and the organization of cell masses into tissues. The extracellular matrix varies in detail according to the type of cell producing it; collagen however is the usual major constituent with variable amounts of proteoglycans and various glycoproteins. The glycoprotein components appear to play important roles in mediating adhesion between cells and extracellular matrices in particular collagen, and hence in regulating cellular metabolism and growth.

There are at least five genetically distinct collagens with rather unique tissue distributions, and specific cell types are usually found to be associated with each (Table 4.2). Several glycoproteins are now identified that mediate cell adhesion to collagen (Table 4.2). These adhesive factors are secreted by cells and at least some of them probably have separate binding sites for collagen and the cell surface. The specificity of these binding sites offer a simple explanation for the preference of various cell types for particular collagens. Some of these factors are also present in serum, normally included in the growth medium for *in vitro* cultivation of cells. The factors may adsorb passively to tissue culture plastic and promote cell attachment and spreading without benefit of a collagen layer.

Cell adhesion to a collagen substratum or plastic surfaces proceeds in three distinct steps: (1) binding of the glycoprotein factor to the plastic or collagen surface; (2) binding of the cells to the surface attached factor;

Table 4.2 Collagens and glycoproteins mediating cell adhesion to collagen

	Collagen type	*Distribution*	*Adhesive glycoprotein*
Interstitial collagens	I	Skin, bone, tendon dentine	Fibronectin
	II	Cartilage, invertebral discs	Chondronectin
	III	Foetal skin, blood vessels, synovial membranes	Fibronectin
Basement membrane collagens	IV	Basement membranes	Laminin, entactin
	AB or V	Smooth muscle, placenta, lung	?

(3) a multiplication of interactions between the cell surface and the glycoprotein molecules, and a reorganization of the cytoskeleton within the cells, particularly microfilaments, with a flattening of the cell cytoplasm onto the substratum (Fig. 4.15). Mitotic cells or cells released from a monolayer culture with trypsin are spheres covered by many folds, blebs and microvilli of various lengths. The longest microvilli or filopodia make the initial contacts upon reattachment to the substratum and evidently carry at their external tips surface molecules that can explore and interact with a suitably adhesive substratum [93]. The filopodia are rich in cytoplasmic microfilaments that form bundles and sheaths as the cells spread and finally act with other cytoskeletal components to fix the stable morphology of the tightly anchored cell.

The steps involved in cell adhesion to substratum take place rapidly (within 1–2 hours) and in the absence of *de novo* protein or nucleic acid biosynthesis, provided divalent cations (Ca^{2+} and Mg^{2+}) are present. In the long term the addition of exogenous adhesive factor can be dispensed with if the cells themselves produce and secrete the factor. Fibroblasts for example require fibronectin to spread out on plastic or collagen in a couple of hours, but after 12–16 hours they have usually produced sufficient fibronectin to promote their own adhesion in the absence of exogenous fibronectin. The role of microfilaments in the spreading process has been shown by the inhibitary effect of cytochalasin B, a drug that prevents polymerization of actin [86].

(a) *Fibronectin.* Fibronectin was the first collagen-binding adhesive glycoprotein to be described (Table 4.3) [94, 95]. In the 1940s John Edsall and his colleagues isolated a blood protein which they called *cold insoluble globulin* or Cig because it was present as a major component of the precipitate that forms when blood is allowed to clot in the cold. Later work has shown that fibronectin has binding affinity for fibrinogen and fibrin which explains its presence in the fibrin clot. Later fibronectin was isolated from the blood and tissues of many vertebrates and the culture media and cell layers of substratum-attached cultures of various avian and mammalian cells, especially cultured fibroblasts. In the intact tissues fibronectin exists predominantly in fibrils and in close apposition to collagen. It is present in basal lamina adjacent to epithelial cell sheets as well as in adjoining connective tissue, in the stroma of lymphatic

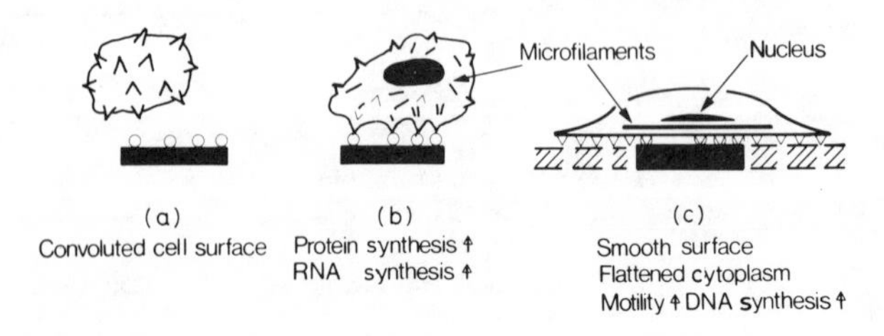

Fig. 4.15 Cell adhesion to substratum. **(a)** A glycoprotein factor binds to the substratum; **(b)** the cells attach; **(c)** and spread.

tissue and around smooth muscle cells and striated muscle fibres. The cellular sites of synthesis of fibronectin in these areas is not well-defined. However, in addition to fibroblasts, myoblasts, astroglial cells, endothelial cells, macrophages, mast cells and epithelial cells in culture produce fibronectin and incorporate some of it into an extracellular matrix often in large quantities.

The circulating form of fibronectin exists as a disulphide-bonded dimer of polypeptide subunits (one somewhat smaller than the other) that are structurally very similar but not identical. The cell-associated form consists of subunits of very similar molecular weight (about 225 000) and may be polymerized by extensive disulphide bonding. However, the dimeric form can be extracted from cells and cell secretions. These various forms of fibronectin appear to be genetically chemically and functionally rather similar but not identical and a simplified general model for the fibronectin molecule is shown in Fig. 4.16. Fibronectin is a glycoprotein containing about four *N*-glycans of structure similar to serum glycoproteins, e.g. orosomucoid. The carbohydrate content is probably not directly involved in biological activity but may protect the molecule from inappropriate proteolysis during

Table 4.3 Properties of fibronectins

Molecular weight	Approx. 450 000
Subunit structure	Dimeric (plasma) Dimeric and higher forms (cells)
Carbohydrate structure	4–5 *N*-glycans
α_2-β electrophoretic mobility	
Immunochemically similar forms	Plasma or cells
Polypeptide similarity	Some interspecies differences
Single gene product in plasma and cells	Probably not
Sites of synthesis	Fibroblasts Astraglial cells Endothelial cells: source of plasma form? Early mesenchymal cells Macrophages Mast cells
Binding domains	Collagens Fibrin (ogen) Heparin Actin DNA Cell surface receptors?
Cross linking	Disulphide bonds Transglutaminase (Factor XIII)
Functions	Mediates adhesion of spreading of cells to collagen Promotes cell motility Prevents differentiation of chondrocytes Acts as a non-specific opsonin in blood Stabilizes blood clotting

synthesis and secretion. As shown dramatically by electron micrographs [96] as well as chemical dissection, the dimeric molecule consists of functionally distinct domains separated by flexible non-globular peptide sequences that are particularly sensitive to proteolysis. Therefore active fragments can be easily obtained and shown to have different biological activities. The 40 000 molecular weight (40K) segment binding to collagen for example is located internally but near the N-terminus of the polypeptide chains. This fragment, like the intact fibronectin, binds to native, fibrillar collagens, in particular during collagen fibril precipitation [97], and better to denatured collagens, of types I–IV. It shows some preference for the insterstial collagens types I and III in keeping with its role in fibroblast adhesion. The peptide sequence of the collagens recognized by the fibronectin 40K domain is a highly conserved region which is recognized also by mammalian collagenase. An early study by Stephen Haushka showed that myoblast differentiation into muscle is supported by exactly the same fragment of collagen and, since fibronectin is synthesized by and mediates the adhesion of myoblasts to collagen, the biological mechanism of the process seems well established in this case.

Fibronectin–collagen binding is relatively weak and is stabilized by a large number of interactions between each cell and the substratum, for example BHK cells make about 50 000 such contacts [86]. Secondly these interactions may be stabilized by secretion from spreading cells of heparan sulphate, a proteoglycan very commonly found to be associated with the underside of cells adherent to a substratum. The main heparan sulphate binding region of fibronectin is located close to the collagen binding region but separate from it.

It is assumed that plasma membrane components are involved in binding to a fibronectin coated surface, and that these communicate directly or indirectly from the outside of the cell into the cytoplasm to trigger the reorganization of microfilaments. Cytoplasmic proteins are known to be present on the inner side of the plasma membrane that precipitate or form anchorage points for the microfilaments. Two proteins in particular have been implicated, namely vinculin, a protein

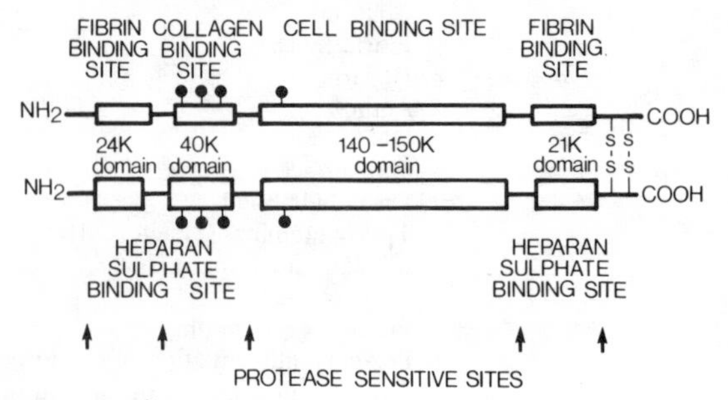

Fig. 4.16 Idealized structure of fibronectin showing the sites of limited proteolysis and the functional domains. ♀, glycans.

of 130 000 molecular weight and α-actinin of 94 000 molecular weight. Antibodies against vinculin appear to stain regions of the underside of substratum-attached cells that do not always contain fibronectin while α-actinin seems to codistribute with fibronectin molecules and is likely to be most intimately involved in the spreading process triggered by a fibronectin-coated surface [98]. The exact nature of the plasma membrane component(s) interacting directly with fibronectin is presently not known. The carbohydrate moieties of glycolipids [94] or glycoproteins [86, 99] have been implicated and possibly more than one factor is involved (Fig. 4.17).

(b) *Laminin.* Although fibronectin can mediate the attachment and spreading of many cells to substrata, it is not always effective. A striking example is the lack of an effect on various epithelial cells that adhere to and spread on collagen layers in the absence of fibronectin [94]. Another glycoprotein, laminin, speeds up the attachment of such cells to collagen and evidently is the fibronectin counterpart for cells of epithelial origin. Laminin is a multimeric glycoprotein molecule of molecular weight 800 000 consisting of subunits of 200 000 and 400 000 molecular weight. Unlike fibronectin, laminin shows high specificity for cells attaching to Type IV collagen, the type present in basement membranes and mostly in contact with epithelial cells in the body. Furthermore, laminin is produced only by cells of epithelial origin. Fibronectin may also be produced by these cells however, and appears to mediate the adhesion of some epithelial cells to interstial collagens *in vitro* where laminin is involved in their adhesion to basement membrane collagen. Laminin, like fibronectin, binds to heparan and heparan sulphate [100]. Since the latter is present in many basement membranes these interactions may play a role in the organization of the matrix and in the adhesion of cells to it.

Other factors recently purified from basal lamina may also play a role in the adhesion of cells to basement membrane collagens. One such factor is entactin, a sulphated glycoprotein of molecular weight 158 000 secreted by endodermal cells in culture and located at the base of the cells facing the basement membrane in intact tissues [101].

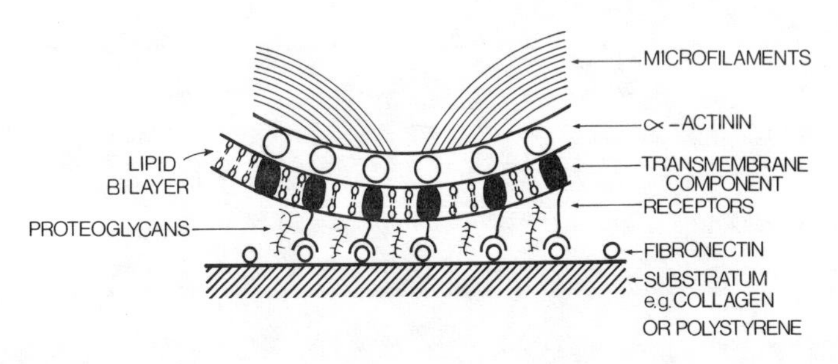

Fig. 4.17 Representation of the molecules involved in cell spreading.

(c) *Chondronectin.* A highly specialized cell, the chondrocyte also requires a unique glycoprotein factor for adhesion to collagen, in this case the specific Type II collagen of cartilage [94]. Chondronectin has a molecular weight of 180 000 and like fibronectin is present in serum. Serum itself can therefore mediate chondrocyte adhesion. Chondronectin binds specifically to collagen Type II characteristic of cartilage. Interestingly, chondrocytes will respond to fibronectin as well as chondronectin. Thus, the cells attached to Type II collagen by chondronectin as cells with rounded morphology, can be induced to spread into fibroblastic forms by addition of fibronectin. The cells then cease to produce cartilage specific collagen and proteoglycans and revert to a fibroblastic phenotype [102]. Fibronectin in this system therefore suppresses a specific cell differentiation event in cell culture, raising the possibility that the production of fibronectin either by the chondrocytes or other cells modulates tissue development. Such a modulation could be important in the formation of myogenic regions within a cartilaginous tissue, and points out the regulatory role of the supporting matrix in directing cell growth and differentiation.

This brief summary of an active area of present research raises a number of outstanding unresolved problems related to glycoprotein function: (1) what are the roles of glycoproteins in regulating the organization of extracellular matrices?; (2) do glycoproteins mediating specific cell adhesion contribute to the unique relationships between differentiated cell types and particular extracellular matrices?; (3) which glycoproteins, glycolipids or proteoglycans associated with cell surfaces are involved in the intimate interactions between cells and the extracellular matrix?; (4) how are these interactions modulated for example to allow cells to move over an adhesive matrix or in malignant cells to migrate from primary sites to infiltrate and colonize other regions of the extracellular matrix?

These and many other questions concerned with glycoproteins are under active investigation and together with the other systems outlined in this book illustrate the widespread interest in the study of the structure and biosynthesis of these important biological macromolecules [103].

Appendix

The following is a brief summary of some important developments that have been reported since completion of the manuscript.

A novel sequence containing an *N*-acetylgalactosamine residue joined glycosidically $\beta 1 \rightarrow 2$ to the $\alpha 1 \rightarrow 6$ linked mannose residue of the trimannosyl $(man)_3$ $(glcNAc)_2$ core region of *N*-glycans, is present in ovine luteinizing hormone [104]. The *N*-acetylgalactosamine unit carries a sulphate group. This is the first well-documented instance of β-linked *N*-acetylgalactosamine residues in *N*-glycans (Section 2.3). A broad specificity *endo-β-N*-acetylglucosaminidase from *Flavobacter meningo-septicum* hydrolyses both oligomannosidic and complex *N*-glycans [105] and will be a useful addition to the collection of enzymes discussed in Section 2.4.

Ankyrin- and spectrin-like molecules have been found [106–108] in several cell types other than erythrocytes, suggesting that these proteins may have a general function in the coupling of actin networks to other cytoskeletal elements, e.g. microtubules, and to membranes in cells (Section 2.6.2).

The pattern of assembly of the major oligosaccharide–lipid intermediate discussed in Section 3.3.1 appears to be followed in mammary gland [109] and in yeasts [110], and the immediate precursor of each α-mannosyl unit has been identified unequivocally [111].

Glucosidase II, the enzyme responsible for removal of $1 \rightarrow 3$ linked glucose residues during processing of *N*-glycans (Section 3.3.2) has been purified [112]. It does not remove the $1 \rightarrow 2$ linked glucose unit which requires another specific glucosidase. A lectin-resistant lymphoma cell line is deficient in glucosidase II, while the cells contain normal activities of glucosidase I [113]. The cells accumulate glucoproteins, strengthening the importance of glucosidases in the formation of 'complex' *N*-glycans. The 'late' α-mannosidase(s) involved in processing the product of *N*-acetylglucosaminyl transferase I action (Section 3.3.3) is powerfully and specifically inhibited by *swainsonine*, an alkaloid isolated from an Australian plant [114]. Some other α-mannosidases are not inhibited. Swainsonine-treated cells fail to produce 'complex' *N*-glycans [115] and ingestion of the plant by animals induces a condition similar to a hereditary lysosomal storage disease, mannosidosis or lysosomal α-mannosidase deficiency, and a chronic neurological disease, locoism [116].

Two of the *N*-acetylglucosaminyl transferases involved in assembly of highly branched N-glycans (Section 3.3.3) have been identified. N-acetylglucosaminyl transferase III which adds an N-acetylglucosamine residue in $\beta 1 \rightarrow 4$ linkage to the β-linked mannose of the trimannosyl core region is present in hen oviduct in high amounts [117], consistent with the preponderance of N-glycans containing this linkage in ovalbumin (Section 2.3.1). The enzyme utilizes either the natural product of N-acetylglucosaminyl transferase I action, glcNAc $(man)_5$ $(glcNAc)_2$, or the product of N-acetylglucosaminyl transferase II action. This last finding explains the existence in many glycoproteins including ovalbumin (see Fig. 2.6) and glycophorin (see Fig. 2.19) of 'complex' N-glycans containing the glcNAc $\beta 1 \rightarrow 4$ man sequence in the core region. The addition of a glcNAc $\beta 1 \rightarrow 4$ unit to the β-mannose residue of the $(man)_5$ $(glcNAc)_2$ structure prevents processing to a three-mannose stage, i.e. prevents formation of an acceptor suitable for the action of N-acetylglucosaminyl transferase II. Hence, during formation of 'complex' N-glycans containing the glcNAc $\beta 1 \rightarrow 4$ man $\beta 1 \rightarrow 4$ glcNAc $\beta 1 \rightarrow 4$ glcNAc sequence, N-acetylglucosaminyl transferase II must act before the transferase III. Another specific enzyme, N-acetylglucosaminyl transferase IV [118] adds an N-acetylglucosaminyl residue in $\beta 1 \rightarrow 4$ linkage to the $\alpha 1 \rightarrow 3$-linked mannose residue of the trimannosyl core region of N-glycans (see Figs 2.6 and 2.11). The preferred substrate is a core region $(man)_3$ $(glcNAc)_2$ sequence substituted on both α-mannosyl units, with N-acetylglucosamine residues in $\beta 1 \rightarrow 2$ linkage. Removal of the N-acetylglucosaminyl units on the mannosyl residue linked $\alpha 1 \rightarrow 6$ to the β-mannose unit, greatly reduces the enzymatic activity, as does substitution of either $\beta 1 \rightarrow 2$ linked N-acetylglucosamine residue with galactose. The substrate specificities of these enzymes, glcNAc transferases III and IV, confirm therefore the exquisite sensitivity of glycosyltransferase towards extended oligosaccharide sequences.

An interesting toxin called *gelonin* is present in the plant *Gelonium multiflorum* [119]. Gelonin appears to be analogous to the toxic subunit A of ricin or abrin (Section 4.2.3). It is ineffective when added to intact cells but is highly toxic to cells when covalently linked to concanavalin A [119] or to specific anti-Thy-1 immunoglobulins [120]. Currently, there is intense interest in the construction of such *chimeric toxins* consisting of the enzymatically active moiety of plant or bacterial toxins linked to a binding moiety, such as a specific antibody or hormone. It is hoped that chimeric toxins may be so designed to have a high cell specificity and be useful in the precise removal of undesired cells, e.g. tumour cells.

A cell surface location for discoidin has been established [121], strengthening proposals (Section 4.5.1) for a role of lectins in slime mold adhesion. The discoidins and pallidins are not integral components of the plasma membrane but are bound to, and therefore bridge, cell surface glycoprotein receptors [122, 123], following the model shown in Fig. 4.12(b). The receptors are expressed at the cell surface throughout the developmental cycle, suggesting strongly that cellular aggregation is controlled by the developmentally-regulated appearance of lectins at the cell surface.

Evidence is increasing for a role of lectins in the adhesion of bacteria to animal tissues as an essential first step in bacterial colonization and pathogenicity [124]. Many Gram-negative bacteria, e.g. *Escherichia coli*, carry surface projections that contain a mannose binding lectin(s) and adhesion of these bacteria to mucosal surfaces can be prevented or reversed with simple mannosides. Although not yet established, it is likely that the natural receptors for these bacterial *adhesin(s)* are oligomannosidic *N*-glycans. The adhesion of other pathogenic bacteria also involves sugar-mediated interactions, e.g. the salivary bacterium *Streptococcus sanguis* which is responsible for dental caries and peridontal disease, binds to glycoproteins containing the sequence neuNAc $\alpha 2 \rightarrow 3$ gal $\beta 1 \rightarrow 3$ galNAc, a common constituent of salivary mucins (Section 2.5.1) [125]. Studies of the mechanisms of bacterial adherence may well lead to the development of effective vaccines and of rather simple (and cheap) sugar inhibitors of bacterial colonization and pathogenicity [126].

A possible role for circulating fibronectin (Cig) (Section 4.5.2) in clearing connective tissue debris is suggested by the finding [127] that gelatin-coated particles are ingested by phagocytes only in the presence of fibronectin and heparin.

The idea that circulating fibronectins and cell-associated fibronectins may be different gene products (Section 4.5.2) is consistent with structural differences shown chemically and using monoclonal antibodies [128–130]. However, the cell-binding domain of fibronectin which has been located [131] in a small proteolytic fragment of molecular weight 12 000 and sequenced, is probably a highly conserved sequence.

References

Inevitably such a short reference list misses out many important contributions. Citation here of any particular paper does not always indicate priority but usually provides the best available access to the literature concerned with the point under discussion. References 1 and 2 are mandatory reading: reference 3 besides containing a fascinating history of glycoproteins is a most valuable source book: other starred references are for general reading.

[1] Ehrlich, P. (1900), *Proc. Roy. Soc. B*, London, **66**, 4246.

[2] Morgan, W.T.J. (1960), *Proc. Roy. Soc. B*, London, **151**, 308–347.

*[3] Gottschalk, A., ed. (1972), *Glycoproteins: their composition structure and function*, Elsevier, Amsterdam.

*[4] Rees, D.A. (1977), *Polysaccharide shapes*, Chapman and Hall, London.

*[5] Brown, R.G. and Kimmins, W.C. (1977), *International Review of Biochemistry II*, Volume 13: *Plant Biochemistry* (Northcote, D.H., ed.), University Park Press, Baltimore, pp. 183–209.

[6] Finne, J., Krusius, T., Margolis, R.K. and Margolis, R.U. (1979), *J. Biol. Chem.*, **254**, 10 295–10 300.

*[7] Montreuil, J. (1979), *Adv. Carb. Chem. Biochem.*, **37**, 157–223.

*[8] Kobata, A. (1979), *Analyt. Biochem.*, **100**, 1–14.

[9] Huang, C.-C., Mayer, H.E. and Montgomery, R. (1970), *Canb. Res.*, **13**, 127–137.

[10] Narasimhan, S., Harpaz, N., Longmore, G., Carver, J.-P., Grey, A.A. and Schachter, H. (1980), *J. Biol. Chem.*, **255**, 4876–4884.

[11] Longmore, G.D. and Schachter, H. (1982), *Carbohydr. Res.*, **100**, 365–392.

[12] Takayasu, T., Takahashi, N. and Shinoda, T. (1980), *Biochem. Biophys. Res. Commun.*, **97**, 635–641.

[13] Chapman, A. and Kornfeld, R. (1979), *J. Biol. Chem.*, **254**, 816–823; 824–828.

[14] Fournet, B., Montreuil, J., Strecker, G., Dorland, L., Haverkamp, J., Vliegenthart, F.G., Binette, J.P. and Schmid, K. (1978), *Biochemistry*, **17**, 5206–5214.

[15] Takahashi, N. and Nishibe, H. (1981), *Biochim. Biophys. Acta*, **657**, 457–467.

[16] Slomiany, B.L., Murty, V.L.N. and Slomiany, A. (1980), *J. Biol. Chem.*, **255**, 9719–9723.

*[17] Hakomori, S-I. (1981), *Seminars in Hematology*, **18**, 39–62.

*[18] Hughes, R.C. (1976), *Membrane glycoproteins: a review of structure and function*, Butterworths, London.

[19] Furthmayr, H. (1978), *J. Supramol. Struct.*, **9**, 79–95.

[20] Yoshima, H., Furthmayr, H. and Kobata, A. (1980), *J. Biol. Chem.*, **255**, 9713–9718.

[21] Fukuda, M. and Fukuda, M. (1981), *J. Supramol. Struct.*, **17**, 313–324.

*[22] Cappuccinelli, P. (1980), *Motility of living cells*, Chapman and Hall, London.

*[23] Parodi, A.J. and Leloir, L. (1979), *Trends Biochem. Sci.*, **4**, 65–67.

*[24] Hughes, R.C. and Butters, T.D. (1981), *Trends Biochem. Sci.*, **6**, 228–230.

[25] Chapman, A., Li, E. and Kornfeld, S. (1979), *J. Biol. Chem.*, **254**, 10 243–10 249.

[26] Spenser, J.P. and Elbein, A.D. (1980), *Proc. Nat. Acad. Sci.*, **77**, 2524–2527.

[27] Trowbridge, I.S. and Hyman, R. (1979), *Cell*, **17**, 503–508.

[28] Bause, E. (1979), *FEBS Lett.*, **103**, 296–270.

[29] Hart, G.W., Brew, K., Grant, G.A., Bradshaw, R.A. and Lennarz, W.J. (1979), *J. Biol. Chem.*, **254**, 9747–9753.

[30] Marshall, R.D. (1974), *Biochem. Soc. Symp.*, **40**, 17–26.

[31] Turco, S.J., Stetson, B. and Robbins, P.W. (1977), *Proc. Nat. Acad. Sci.*, **74**, 4411–4414.

[32] Spiro, R.G., Spiro, M.J. and Bhoyroo, V.D. (1979), *J. Biol. Chem.*, **254**, 7659–7667.

[33] Tabas, I. and Kornfeld, S. (1978), *J. Biol. Chem.*, **253**, 7779–7786.

[34] Li, E. and Kornfeld, S. (1979), *J. Biol. Chem.*, **254**, 1600–1605.

*[35] Schachter, H., Narasimhan, S. and Wilson, J.R. (1978), in *Glycoproteins and Glycolipids in Disease Processes* (Walborg, E.J., ed.), American Chemical Society, Washington D.C., pp. 21–46.

[36] Sheares, B.T., Lau, J.J.Y. and Carlson, D.M. (1982), *J. Biol. Chem.*, **257**, 599–602.

[37] Fukuda, M.N., Papermaster, D.S. and Hargrave, P.A. (1979), *J. Biol. Chem.*, **254**, 8201–8207.

[38] Paulson, J.C., Prieels, J.P., Glasgow, L.R. and Hill, R.L. (1978), *J. Biol. Chem.*, **253**, 5617–5624.

[39] Yamashita, K., Tachibana, Y., Nakayama, T., Kitamura, M., Endo, Y. and Kobata, A. (1980), *J. Biol. Chem.*, **255**, 5635–5642.

*[40] Bailey, A.J. and Robbins, S.P. (1976), *Sci. Prog. Oxford*, **63**, 419–444.

[41] Young, J.D., Tsuchiya, D., Sandlin, D.E. and Holroyde, M.J. (1979), *Biochemistry*, **18**, 4444–4448.

[42] Williams, D. and Schachter, H. (1980), *J. Biol. Chem.*, **255**, 11 247–11 252.

[43] Sadler, J.E., Rearick, J.I., Paulson, J.C. and Hill, R.L. (1979), *J. Biol. Chem.*, **254**, 4434–4443.

[44] Bergh, M.L.E., Hooghwinkel, G.J.M. and Eijnden, D.H. van den (1981), *Biochim. Biophys. Acta*, **660**, 161–169.

[45] Berger, E.G., Mandel, T. and Schilt, V. (1981), *J. Histochem. Cytochem.*, **29**, 364–370.

[46] Blobel, G. and Dobberstein, B. (1975), *J. Cell Biol.*, **67**, 835–851; 852–862.

[47] Das, R.C. and Heath, E.C. (1980), *Proc. Nat. Acad. Sci.*, **77**, 3811–3815.

[48] Rothman, J.E. and Fine, R.E. (1980), *Proc. Nat. Acad. Sci.*, **77**, 780–784.

[49] Hanover, J.A., Lennarz, W.J. and Young, J.D. (1980), *J. Biol. Chem.*, **255**, 6713–6716.

[50] Mills, J.T. and Adamany, A.M. (1978), *J. Biol. Chem.*, **253**, 5270–5273.

[51] Carson, D.D. and Lennarz, W.J. (1981), *J. Biol. Chem.*, **256**, 4679–4686.
[52] Tavares, I.A., Coolbear, T. and Hemming, F.W. (1981), *Arch. Biochem. Biophys.*, **207**, 427–436.
*[53] Watkins, W.M. (1980), *Adv. Human Genetics*, **10**, 1–136.
[54] Crine, P., Seidah, N.G., Routhier, R., Gossard, F. and Chretien, M. (1980), *Eur. J. Biochem.*, **110**, 387–396.
[55] Sharon, N. and Lis, H. (1977), In *The Antigens*, Vol. 4 (Sela, M., ed.), Academic Press, New York, pp. 429–463.
[56] Baenziger, J.U. and Fiete, D. (1979), *J. Biol. Chem.*, **254**, 2400–2407.
*[57] Olsnes, S. and Pihl, A. (1976), in *Receptors and Recognition B1: The specificity and action of animal, bacterial and plant toxins* (Cuatrecasas, P., ed.), Chapman and Hall, London, pp. 129–173.
[58] Baenziger, J.U. and Fiete, D. (1979), *J. Biol. Chem.*, **254**, 9795–9799.
[59] Vischer, P. and Hughes, R.C. (1981), *Eur. J. Biochem.*, **117**, 275–284.
[60] Hunt, L.A. (1980), *J. Virol.*, **35**, 362–370.
[61] Chapman, A., Fujimoto, K. and Kornfeld, S. (1980), *J. Biol. Chem.*, **255**, 4441–4446.
[62] Trowbridge, I.S. and Hyman, R. (1978), *Eur. J. Immunol.*, **8**, 716–723.
[63] Hercz, A., Katona, E., Cutz, E., Wilson, J.R. and Barton, M. (1978), *Science*, **201**, 1229–1232.
*[64] Neufeld, E. and Ashwell, G. (1980), in *The Biochemistry of Glycoproteins and Proteoglycans* (Lennarz, W.J., ed.), Plenum Press, New York, pp. 252–257.
[65] Varki, A. and Kornfeld, S. (1980), *J. Biol. Chem.*, **255**, 10 847–10 858.
[66] Fischer, H.D., Gonzalez-Noriega, A. and Sly, W.S. (1980), *J. Biol. Chem.*, **255**, 5069–5074.
[67] Reitman, M.L. and Kornfeld, S. (1981), *J. Biol. Chem.*, **256**, 4275–4281.
[68] Waheed, A., Pohlman, R., Hasilik, A. and von Figura, K. (1981), *J. Biol. Chem.*, **256**, 4150–4152.
*[69] Dean, R.T. (1978), *Cellular degradative processes*, Chapman and Hall, London.
*[70] Ashwell, G. and Morell, A.G. (1977), *Trends Biochem. Sci.*, **2**, 76–78.
[71] Baenziger, J.U. and Fiete, D. (1980), *Cell*, **22**, 611–620.
[72] Hatton, M.W.C., März, L., Berry, L.R., Debanne, M.I. and Regoeczi, E. (1979), *Biochem. J.*, **181**, 633–638.
[73] Stockert, R.J., Gartner, U., Morell, A.C. and Wolkoff, A.W. (1980), *J. Biol. Chem.*, **255**, 3830–3831.
[74] Steer, C.J. and Ashwell, G. (1980), *J. Biol. Chem.*, **255**, 3008–3013.
[75] Stahl, P.O., Schlesinger, P.H., Rodman, J.S. and Doebber, T. (1976), *Nature*, **264**, 86–88.
[76] Mizuno, Y., Isozutsumi, Y., Kawasaki, T. and Yamashina, I. (1981), *J. Biol. Chem.*, **256**, 4247–4252.
[77] Warr, G.A. (1980), *Biochem. Biophys. Res. Commun.*, **93**, 737–745.
[78] Drickamer, K. (1981), *J. Biol. Chem.*, **256**, 5827–5839.
[79] Aminoff, D., Bell, W.C. and Vorder Bruegge, W.G. (1978), in *Cell Surface Carbohydrates and Biological Recognition* (Marchesi, V.T., Ginsburg, V., Robbins, P.W. and Fox, C.F., eds), Alan R. Liss, New York, pp. 569–581.
[80] Kolb, H. and Kolb-Bachofen, V. (1978), *Biochem. Biophys. Res. Commun.*, **85**, 678–683.

[81] Hildenbrandt and Aronson, N.N. (1979), *Biochim. Biophys. Acta.*, **587**, 373–380.
*[82] Burger, M.M. and Jumblatt, J. (1977), in *Cell and Tissue Interactions* (Lash, J.W. and Burger, M.M., eds), Raven Press, New York, pp. 155–172.
[83] Muller, W.E.G., Zahn, R.K., Kuretec, B., Miller, I. and Uhlenbruck, G. (1979), *J. Biol. Chem.*, **254**, 1280–1287.
*[84] Moscona, A.A. and Hausman, R.E. (1977), in *Cell and Tissue Interactions* (Lash, J.W. and Burger, M.M., eds), Raven Press, New York, pp. 173–185.
[85] Oppenheimer, S.B. (1975), *Exptl. Cell Res.*, **92**, 122–126.
*[86] Hughes, R.C., Pena, S.D.J. and Vischer, P. (1979), in *Cell Adhesion and Motility* (Curtis, A.S.G. and Pitts, J.D., eds), Cambridge University Press, pp. 329–356.
*[87] Roseman, S. (1970), *Chem. Phys. Lipids*, **5**, 270–278.
[88] Rauvala, H. and Hakomori, S-I. (1981), *J. Cell Biol.*, **88**, 149–159.
*[89] Barondes, S.H. (1980), in *Cell Adhesion and Motility* (Curtis, A.S.G. and Pitts, J.D., eds), Cambridge University Press, pp. 309–324.
[90] Harrison, F.J. and Chesterton, C.J. (1980), *Nature*, **286**, 502–504.
[91] Childs, R.A. and Feizi, T. (1979), *FEBS Lett.*, **99**, 175–179.
[92] Simpson, D.L., Thorne, D.R. and Loh, H.H. (1977), *Nature*, **266**, 367–369.
*[93] Grinnell, F. (1978), *Int. Rev. Cytol.*, **58**, 65–144.
*[94] Kleinman, H.K., Klebe, R.J. and Martin, G.R. (1981), *J. Cell Biol.*, **88**, 473–485.
*[95] Ruoslahti, E., Engvall, E. and Hayman, E. (1981), *Coll. Res.*, **1**, 95–128.
[96] Engel, J., Odermatt, E., Engel, A., Madri, J.A., Furthmayr, H., Rhode, H. and Timpl, R. (1981), *J. Mol. Biol.*, **150**, 97–120.
[97] Kleinman, H.K., Wilkes, C.M. and Martin, G.R. (1981), *Biochemistry*, **20**, 2325–2330.
[98] Chen, W.T. and Singer, S.J. (1980), *Proc. Nat. Acad. Sci.*, **77**, 7318–7322.
[99] Aplin, R.C., Hughes, R.C., Jaffe, C.L. and Sharon, N. (1981), *Exptl. Cell Res.*, **134**, 488–494.
[100] Sakashita, S., Engvall, E. and Ruoslahti, E. (1980), *FEBS Lett.*, **116**, 243–246.
[101] Carlin, B., Jaffe, R., Bender, B. and Chung, A.E. (1981), *J. Biol. Chem.*, **256**, 5209–5214.
[102] West, C., Lanza, W.R., Rosenbloom, J., Lowe, M., Holtzer, H. and Avdavolic, N. (1979), *Cell*, **17**, 491–501.
*[103] Hughes, R.C. and Pena, S.D.J. (1981), in *Carbohydrate Metabolism and Its Disorders*, Vol. 3 (Randle, P.J., Steiner, D.F. and Whelan, W.J., eds), Academic Press, New York, pp. 363–423.
[104] Bedi, G.S., French, W.C. and Bahl, O. (1982), *J. Biol. Chem.*, **257**, 4345–4355.
[105] Elder, J.H. and Alexander, S. (1983), *Proc. Natl. Acad. Sci. USA*, **79**, 4540–4544.
[106] Bennett, V. and Davis, J. (1981), *Proc. Natl. Acad. Sci. USA*, **78**, 7550–7554.
[107] Bennett, V., Davis, J. and Fowler, W.E. (1982), *Nature*, **299**, 126–130.
[108] Glenney, J., Glenney, P. and Weber, K. (1982), *Cell*, **28**, 843–854.

[109] Vijay, I.K. and Perdew, G.H. (1982), *Eur. J. Biochem.*, **126**, 167–172.
[110] Prakash, C. and Vijay, I.K. (1982), *Biochemistry*, **21**, 4810–4818.
[111] Rearick, J.I., Fujimoto, K. and Kornfeld, S. (1981), *J. Biol. Chem.*, **256**, 3762–3769.
[112] Burns, D.M. and Touster, O. (1982), *J. Biol. Chem.*, **257**, 9991–10 000.
[113] Reitman, M.L., Trowbridge, I.S. and Kornfeld, S. (1982), *J. Biol. Chem.*, **257**, 10 357–10 363.
[114] Tulsiani, D.R.P., Harris, T.M. and Touster, O. (1982), *J. Biol. Chem.*, **257**, 7936–7939.
[115] Elbein, A.D., Dorling, P., Vosbeck, K. and Horisberger, M. (1982), *J. Biol. Chem.*, **257**, 1573–1576.
[116] Molyneaux, R.J. and James, L.F. (1982), *Science*, **216**, 190–191.
[117] Narasimhan, S. (1982), *J. Biol. Chem.*, **257**, 10 235–10 242.
[118] Gleeson, P., Vella, G., Narasimhan, S. and Schachter, H. (1982), *Fed. Proc.*, **41**, 1147.
[119] Stirpe, F., Olsnes, S. and Pihl, A. (1980), *J. Biol. Chem.*, **255**, 6947–6953.
[120] Thorpe, P.E., Brown, A.N.F., Ross, W.C.J., Cumber, A.J., Detre, S.I., Edwards, D.C., Davies, A.J.S. and Stirpe, F. (1981), *Eur. J. Biochem.*, **116**, 447–454.
[121] Madley, I.C., Cook, M.J. and Hames, B.D. (1982), *Biochem. J.*, **204**, 787–794.
[122] Ray, J. and Lerner, R.A. (1982), *Cell*, **28**, 91–98.
[123] Drake, D.K. and Rosen, S.D. (1982), *J. Cell Biol.*, **93**, 383–389.
[124] Elliott, K., O'Conner, M. and Whelan, J. (eds) (1982), *CIBA Symposium 80. Adhesion and micro-organism pathogenicity.* Pitman Medical, London.
[125] Murray, P.A., Levine, M.J., Tabak, L.A. and Reddy, M.S. (1982), *Biochem. Biophys. Res. Commun.*, **106**, 390–396.
[126] Aronson, M., Medalia, O., Schori, L., Mirelman, D., Sharon, N. and Ofek, I. (1979), *J. Infect. Dis.*, **139**, 329–332.
[127] Van de Water, L., Schroeder, S., Crenshaw, E.B. and Hynes, R.O. (1981), *J. Cell Biol.*, **90**, 32–39.
[128] Fukunda, M., Levery, S.B. and Hakomori, S.I. (1982), *J. Biol. Chem.*, **257**, 6856–6860.
[129] Hayashi, M. and Yamada, K.M. (1981), *J. Biol. Chem.*, **256**, 11 292–11 300.
[130] Atherton, B.T. and Hynes, R.O. (1981), *Cell*, **25**, 133–141.
[131] Pierschbacher, M.D., Ruoslahti, E., Sundelin, J., Lind, P. and Peterson, P.A. (1982), *J. Biol. Chem.*, **257**, 9593–9597.

Index